CW01347427

Network Systems Design

Network Systems Design

Edited by

Erol Gelenbe
Duke University
Durham, North Carolina

Kallol K. Bagchi
Florida Atlantic University at Boca Raton

and

George W. Zobrist
University of Missouri at Rolla

Foreword by Darrell D. E. Long

Gordon and Breach Science Publishers
Australia Canada China France Germany India Japan Luxembourg
Malaysia The Netherlands Russia Singapore Switzerland

Copyright © 1999 OPA (Overseas Publishers Association) N.V. Published by license under the Gordon and Breach Science Publishers imprint.

All rights reserved.

No part of this book may be reproduced or utilized in any form or by any means, electronic or mechanical, including photocopying and recording, or by any information storage or retrieval system, without permission in writing from the publisher. Printed in Malaysia.

Amsteldijk 166
1st Floor
1079 LH Amsterdam
The Netherlands

British Library Cataloguing in Publication Data

Network systems design
  1. Data transmission systems 2. System design 3. Computer network architecture
  I. Gelenbe, Erol II. Bagchi, Kallol K. III. Zobrist, George W. (George Winston), 1934 -
  004.2'1

ISBN 90-5699-635-5

*This book is dedicated to
all the people
who helped in making it a success.*

Editorial Board   *Onno Boxma, Marco Ajmone Marsan,
Darrell D. E. Long, Jason Lin* and
Pawel Gburzynski

CONTENTS

Foreword   ix

Editors' Preface   xi

Acknowledgments   xiii

1 Wormhole Data Routing in Multiprocessors: Performance Simulation   1
   Yen-Wen Lu

2 Modeling and Simulation of a Communication Protocol by Stochastic Well-Formed Nets   33
   Rossano Gaeta, Matteo Sereno and Giovanni Chiola

3 Modeling and Analysis of Multi-Access Mechanisms in SUPERLAN   47
   Adrian Popescu and Rassul Ayani

4 Modeling and Management of Self-Similar Traffic Flows in High-Speed Networks   69
   Ashok Erramilli, Walter Willinger and Jonathan L. Wang

5 A New Traffic Control Mechanism for Continuous Media Communications   97
   Frank Ball, David Hutchison and Demetres Kouvatsos

6 Simulation Modeling of Local and Metropolitan Area Networks   117
   Marco Conti and Lorenzo Donatiello

7 An Improved Model of Heterogeneous Elevator (SCAN) Polling   143
   Michael S. Borella and Biswanath Mukherjee

8 Simulation Modeling of Weak-Consistency Protocols   161
   Richard A. Golding and Darrell D. E. Long

9	Modeling a Multimedia System for VOD Services *Antonio Puliafito, Salvatore Riccobene,* *Giancarlo Iannizzotto and Lorenzo Vita*	187
10	Modeling, Simulation and Synthesis of High-Performance ATM Protocols and Multimedia Systems *Georg Carle and Jochen Schiller*	203
11	Modeling Techniques for PCS Networks *Yi-Bing Lin*	223
12	The Effect of Request-Waiting in a Mobile Computing Network *Yi-Bing Lin and Wai Chen*	241
Notes on Contributors		261
Author Index		271
Subject Index		273

FOREWORD

This volume, *Network Systems Design*, concerns the modeling and simulation of some of the most complex engineering artifacts ever created. In particular, it explores modeling and simulation techniques for communication networks and the protocols they employ. These networks range from the internal network of a multiprocessor, to the global Internet. Construction of these systems is complex and expensive; we can no longer attempt to build them and hope they will function properly. Costs can be in the billions of dollars, and few companies or governments can afford a failure of that magnitude. Similarly, while prototypes are useful, and in many cases essential, the scale of an increasing number of these systems is such that small prototypes may not reveal the true dynamic behavior of the system. At one time, bridges and cathedrals were built through trial and error. If they fell down, the builders would try again, perhaps using a larger stone or a steeper arch. Later, when the mathematics were better understood but still intractable, design rules and approximations were used. Today, we can construct highly accurate models of structures, wings, computers and communication systems that give us a high degree of confidence they will function as designed. A quote from Boulle's *The Bridge over the River Kwai*[*] illustrates the usefulness of modeling before construction begins:

Mechanics, in fact, entail a complete a priori knowledge; and this mental creation, which precedes the material creation, is not the least important of the many achievements of the Western genius.

While Boulle may be inaccurate in characterizing modeling as a strictly Western invention, his comment does emphasize its importance. Few objects in our modern world are made without first being modeled. Modeling can be accomplished with analytical models, continuous or discrete event simulation, or a combination of these techniques. Analytical modeling and discrete event simulation are particularly well-suited for computers and communication systems. Given its wide applicability, it is surprising that modeling receives relatively little attention in most computer science and engineering curricula. Perhaps this series will help to correct this omission.

---

[*] Pierre Boulle, *The Bridge over the River Kwai*, translated by X. Fielding, New York: Vanguard Press, 1954, p. 89.

In this volume there are chapters covering the state of the art in analytical modeling and simulation. Topics include the interconnect of a multiprocessor, local, metropolitan and wide area networks, protocols, multimedia, video-on-demand, wide area applications and telephony. Techniques include Markov modeling, stochastic networks, Monte Carlo and discrete event simulation. The book will interest practitioners, researchers and students.

Darrell D. E. Long

EDITORS' PREFACE

Performance modeling and simulation have been used extensively in computer and communication system design for some time. This book makes the argument that this topic has become a central issue in computer science and engineering research and should be considered seriously. Its object is to lead researchers, practitioners and students involved in this discipline. Distinctive in many ways, it provides tutorials and surveys on important topics, relates new research results that should be of interest to all working in the field and covers a broad area in the process. Each chapter presents background, describes and analyzes important work in the field and provides direction to the reader on future work and further readings. The volume can be used as a reference book by all associated with computers and communication networks. Our hope is that this set of carefully selected papers will be of interest to all computer and communication scientists and engineers (students, academicians, practitioners) in general.

This is the second book of a two-book set on performance modeling and simulation of networks. It deals primarily with applications and systems related to network design. Its companion volume describes theory, tools and techniques for various such systems.

The book begins with a chapter on wormhole data routing simulation and performance. In the first volume, this author had a tutorial on wormhole data routing. Here, a few routing algorithms, adaptive and deterministic, are introduced and tested. Adaptive algorithms usually performed better. The author also obtained better performance from systems with more virtual channels. The second chapter is on Petri net based modeling of communication protocols. The authors describe a new formalism of Petri nets, called Timed Well-formed Nets, and use it to describe a ring communication protocol. Exploiting the model symmetry, this formalism significantly reduces the simulation event-list length, and the simulation is thus kept manageable.

In chapter 3, on multi-access modeling of SUPERLANs, SUPERLAN is defined as an integrated multi-Gbps LAN with data rates of 9.6 Gbps capacity per channel. A class of Media Access Control (MAC) protocols is modeled and some performance analyses are presented. Chapter 4 covers modeling of self-similar traffic in high-speed networks. Impact of traffic on traffic management for high-speed networks is reviewed, and it is shown that self-similar traffic models are best to describe complex traffic patterns and techniques to generate self-similar traffic.

The next chapter (chapter 5) discusses a new traffic control mechanism modeling for ATM- and FDDI-like networks. This mechanism matches temporal aspects of continuous network traffics to temporal network structures. Performance results from such mapping of a video source onto an FDDI network are included. Chapter 6 deals with simulation techniques used in LAN/MAN protocol modeling. Techniques for modeling DQDB, Variable Bit Rate Video and wireless LAN protocols are described.

The seventh chapter covers SCAN polling methods. An exact analysis for finite-buffer heterogeneous elevator polling is introduced. The unique SAN (Stochastic Activity Network)-based technique used in the article is general in nature and has applications in performance in multiple channel networks. Chapter 8 discusses weak-consistency replication protocol simulation modeling. This type of protocol is suitable for mobile networks, robust and fault-tolerant. The authors include detailed simulation analysis of the protocol.

In the ninth chapter, on Petri net based modeling of Video on Demand (VOD) systems, an algorithm for I/O subsystems—based on disk array technology—to manage continuous media data for VOD services is described and modeled. In chapter 10, on an ATM protocol for multi-point error control, the authors include design flow for implementation and performance results. A network element called Group Communication Server allows combination of different error control schemes and a hierarchical approach to multi-cast error control.

Chapter 11 presents realistic analytical models of large-scale Personal Communications Services (PCS) networks. Models for portable population and movement are introduced. Simulation may not be cost-effective for these applications. The final chapter describes a superior channel allocation algorithm in a mobile computer environment. A buffering channel allocation algorithm called Request-Waiting is introduced. The mechanism is shown to be more performance-effective than the user re-dialing method.

ACKNOWLEDGMENTS

Dr. Carey Williamson of the University of Saskatchewan in Canada and Prof. Yi-Bing Lin of National Chiao Tung University in Taiwan deserve special mention for providing help with LaTeX script, which many authors used in preparing final versions of their chapters. It was a pleasure to work with Frank Cerra and his associate Tara Lynch. Tara provided enormous support in the final stages of the book.

Many authors provided feedback for improving the structure of the book. Thanks are due to all of them. We are grateful to the reviewers, editorial board members and all the contributors, who deserve a great deal more than acknowledgment. Without their help, patience and cooperation, this book would not have been possible.

CHAPTER 1

WORMHOLE DATA ROUTING IN MULTIPROCESSORS: PERFORMANCE SIMULATION

Yen-Wen Lu

1. INTRODUCTION

Parallel processing provides significant computational advantages for many scientific, signal processing, and image processing applications. However, as technology and processing power continue to improve, inter-processor communication becomes a performance bottleneck. An important issue for improving the performance of multiprocessor communication is how to utilize channel bandwidth efficiently. Virtual channels were introduced by Dally to increase network capacity and throughput[4]. A physical channel is shared by several virtual channels which are time-multiplexed on the same link. Many deadlock-free algorithms are based on the virtual channel assignment to prevent cyclic dependency in the data flow graph[6,9,10]. Furthermore, communication links can be organized as a pair of opposite uni-directional channels, or combined into a single bi-directional channel between two adjacent nodes. Different link configurations will affect the efficiency of channel utilization, latency, and throughput. Some qualitative comparisons of link configurations have been stated in literature[11,14], but no systematic study has been conducted. In this chapter, we will compare the performance of a single bi-directional channel and two uni-directional channels quantitatively based on constant total channel width, and discuss their trade-offs and practical issues.

A wormhole routing simulator was built to study different routing algorithms and some design tradeoffs: for example, buffer size, virtual channel numbers, channel arbitration, and so on. We are most interested in k-ary n-cube networks because of their generality, regularity, and simplicity. Most of the network simulations assume it takes only a single cycle for a hop[1,2,5,8,9,15], which is not an accurate model for hardware implementation and cannot take care of variable propagation delay. Therefore, we have implemented four stages of pipeline in the router model to simulate a realistic hardware design. In this chapter, we will describe the simulator models and architectures, and some different traffic models. We will also describe the results of the simulation and interpret these in terms of design issues.

1.2. SIMULATION MODELS

The wormhole network simulator can be divided into three levels: the network, the node, and the link model. The network model instantiates the network topology and interconnections, and defines individual nodes in the network. The node model defines the functions inside a router that perform data routing. The link model implements the low level data transfer, interconnection protocols, and FIFO buffers. In addition to the above three level models, we also have a traffic model which generates packets into the network based on the specified injection model and destination distribution. We will describe the details of each model in the following sections.

1.2.1. Network Model

The network model defines the network interconnection. Network dimension and size are declared in this level. A network configuration file contains the declaration of nodes, each port of every node, and the connection with other nodes. We can have an arbitrary network topology by specifying the interconnections in the network configuration file. The network interconnection consistency is checked automatically after reading the configuration file to make sure all the interconnections are legal and one-to-one. Two kinds of nodes can be declared: routers and hosts. The router node will be described by the node model, and the host will be described by the traffic model.

We are most interested in k-ary n-cube networks because of their generality, regularity, and simplicity. A k-ary n-cube mesh network is an n-dimensional grid consisting of k^n nodes. There are k nodes in each dimension, and each node is connected to its Cartesian neighbors (The node here contains a router and a host.). A torus is a mesh with wrap-around connections, *i.e.*, there is a ring in every dimension. With the physical limitation of wire density and bisection width, Dally has shown that low-dimensional cubes perform better than high-dimensional cubes in parallel processing[3]. Therefore, we will concentrate on the two-dimensional mesh or torus in our simulation.

1.2.2. Node Model

The node model defines the router functions and performs data routing. Fig. 1.1 shows the router architecture proposed and implemented in the simulator. A router is composed of header decoders, requesting units, arbiters, crossbar switches, I/O buffers, and I/O controllers. The I/O buffers and I/O controllers will be described in the link model. Each function in a node is declared as a simulation module and the interface between modules is well defined. In this way, we can replace a module and implement different algorithms for a particular function.

In order to enhance the routing efficiency, each packet has the directional information in its header to indicate the direction (positive or negative) it will take in any dimension. We need n bits for an n dimensional array. The directional information is encoded at the source when the packet is generated, and is never changed on the path to the destination. There are n arrival bits in the header associated with each dimension to indicate whether this dimension needs further correction or not. There are other n wrap-around bits in the header to keep track of which dimensions have taken the wrap-around connections. So there are a total of $3n$ bits of supplement information encoded in the header flit of a packet.

A credit feedback scheme is used for traffic congestion control. We assign a credit to each output-input virtual channel pair. The credit is used to indicate the buffer availability at the input port of the receiver and is kept track of at the output port of the sender. The initial value of a credit is equal to the virtual channel buffer size. When a flit is transferred to the output port, the credit of the virtual channel it will use is decremented by one. When the receiving node reads a flit out of

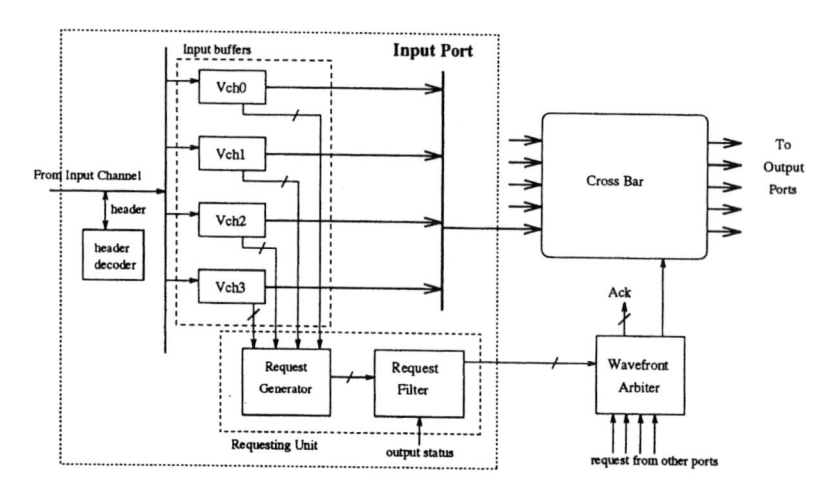

Figure 1.1: Router node architecture model

the corresponding input virtual channel buffers, it will send a credit back to the sender's output port. The credit for the virtual channel at the output port is incremented by one when a credit is received. Because of the propagation delay, the credit at the output port is always less than or equal to the available buffer size of the input port at the receiver. Thus we will not overflow the input buffer and never discard a packet.

We now describe the details of each functional block in a router.

Header Decoder The header decoder at each input port compares the coordinates of the current node and the incoming flit's destination if it is a header. We only have to check the address coordinate in the dimension from which the packet is coming. The header decoder marks the arrival bit of this dimension as "arrived" if the incoming packet's destination matches the current node in this dimension. If the input link is a wrap-around connection, the header decoder will set the wrap-around bit in this dimension for the incoming packet.

Requesting Unit There are two sub-blocks in the requesting unit: a request generator and a request filter. The request generator takes the requests from all virtual channels of the input port and selects the output port and output virtual channel for each request. If the request is issued by a header, the request generator will choose the

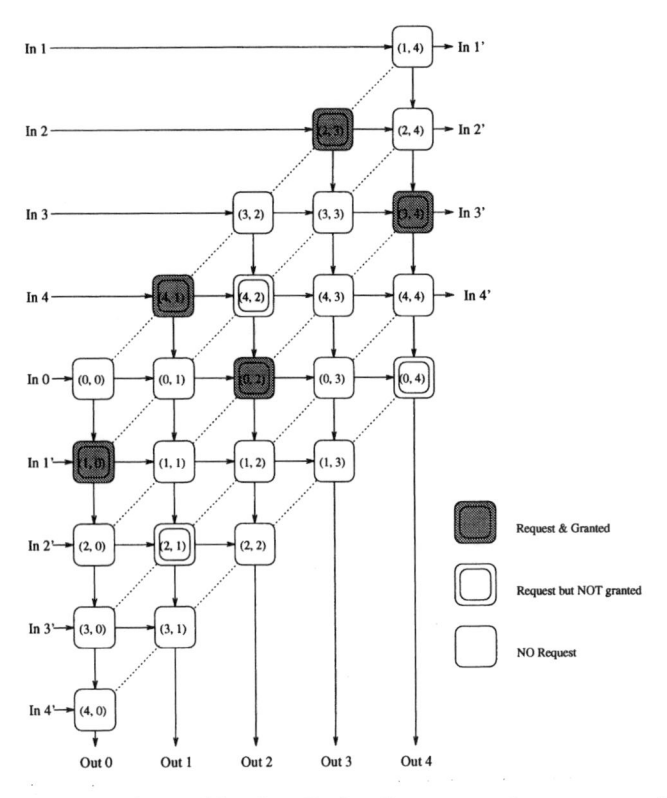

Figure 1.2: Wavefront arbitration. Each cell corresponds to a cross point in the crossbar switch.

output virtual channel based on the routing algorithm for deadlock avoidance and credit availability. If it is not a header issuing the request, then the previous output virtual channel assigned for the same packet is chosen. Before the requests go to the arbiter, they go to the request filter which can block some requests. A request is blocked if (1) the chosen output virtual channel has already had another message in progress, (2) the chosen output virtual channel is full (no credit), or (3) the chosen output port buffer is full. We rule out the requests which satisfy any of the above conditions so they will not occupy the slot for arbitration.

Arbiter The arbiter prevents input and output conflicts at the crossbar and makes switch utilization and throughput as high as possible. We implemented wavefront arbitration to optimize the utilization of the crossbar switch[16]. Fig. 1.2 shows a 5 × 5 wavefront arbiter. Each cell receives the request and generates the grant for the corresponding cross point in the crossbar switch. The cells are arranged in diagonal lines, *i.e.*, wavefronts. Since the requests on a wavefront are from different input ports and for different output ports, there is no conflict between the arbitration cells on the same wavefront. The order of wavefronts determines the priority of requests. The arbitration begins from the top wavefront which has the highest priority. If a request is granted, the corresponding input and output port are disabled and no other requests can be granted for either the same input or output port. To guarantee fair arbitration, we can sort the wavefronts by the "age" of the flits issuing the requests. A more practical compromise is to put the oldest flits on the top wavefront, and the other requests are ordered in a random or round-robin priority.

Crossbar Switch The crossbar switch sets up physical connections from input ports to other output ports. The size of crossbar switch is $m \times m$, where m is the number of ports.

In a two-dimensional mesh or torus, each node has five I/O ports: four to the nearest neighbors and one to the host. The crossbar is a 5 × 5 fully-connected switch such that any input port can be connected to any other output port as long as there is no output conflict.

1.2.3. Link Model

The link model includes the physical channels for data transfer, I/O controller for interconnection protocols, and I/O buffers for virtual channels.

1.2.3.1. Channel Configuration

The links between two adjacent nodes can be organized as a pair of opposite uni-directional channels, one for transmitting and the other for receiving, or combined into a single bi-directional channel (Fig 1.3). Different link configurations will affect the efficiency of channel utilization.

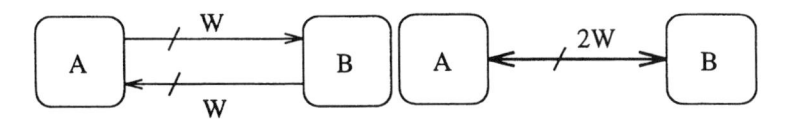

(a) Two uni-directional channels   (b) Single bi-directional channel

Figure 1.3: Channel configuration

If the total link width is a constant $2W$, then a single bi-directional channel can have full channel width $2W$, but each uni-directional channel can only have half of the channel width. When two uni-directional channels are not fully utilized, one may be busy while the other is idle, and half of the channel bandwidth is wasted. However, bi-directional channels may have longer propagation delay due to more capacitance loading on the channel. A special arbitration is necessary for bi-directional channels to prevent conflict and deadlock, and this arbitration will introduce some overhead. When we have more bandwidth to transmit data, the packet length becomes shorter in terms of flits because each flit size is doubled and the same amount of information can be encoded into half of the number of flits. The buffer storage is also halved in terms of the number of flits if the total storage space is constant. Therefore if a packet is blocked in the network, it will distribute in the same number of nodes no matter which scheme we use.

A token exchange mechanism is used for the single bi-directional channel to prevent conflict caused by two neighboring nodes using the channel at the same time. There is a token associated with each physical channel. Only the node with the token can use the channel to transmit data. A node without the token can only listen to the channel as a receiver. When a receiving node wants to transmit data, it has to send a request to the other node which owns the token currently. The current owner can grant the request by sending an acknowledgment back to the requesting node, and the roles of these two nodes exchange[7].

The state diagram of the token exchange is shown in Fig. 1.4. Two neighboring nodes, A and B, start at state *No_token* and *With_token*, respectively. The signal TE is low initially. Node A, which does not

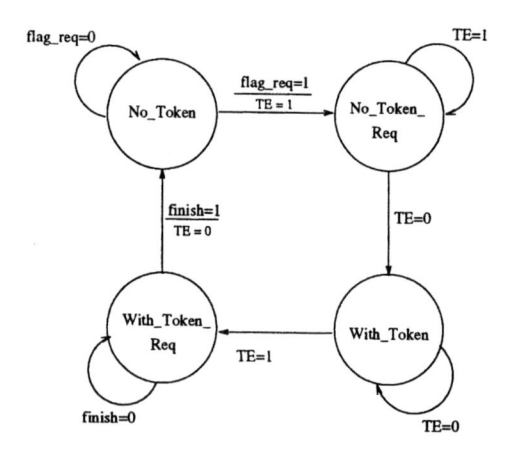

Figure 1.4: Token exchange state diagram

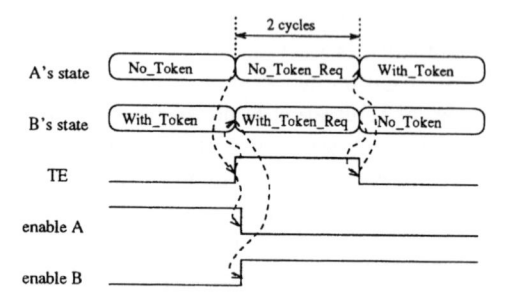

Figure 1.5: Token exchange timing diagram

have the token, can send a request by making *TE=1*, and enter state *No_token_req*. Node B, which is in state *With_token*, goes to state *With_token_req* after seeing *TE* become high. Node B will wait in state *With_token_req* until the *finish* signal goes high, which means it can give up the token now. Then node B will lower *TE* and enters state *No_token*. Node A, which now is waiting in state *No_token_req*, senses the signal *TE* becoming low, and knows that the token has been granted. It enters state *With_token* and begins to transmit data. Fig. 1.5 shows the timing diagram of the token-exchange handshaking between node A and B where *enableA* and *enableB* designate which node is driving the signal *TE*.

One of two conditions must obtain if a node is to give up its token

($finish=1$): 1) if it is idle in the previous cycle, or 2) if the previous flit it sent is a tail of a packet.

Theorem 1 *The token-exchange channel arbitration is conflict-free and deadlock-free.*

Proof: First, we will prove the property of "conservation of token", *i.e.*, there is one and only one token associated with a channel all the time. From the definition of token states, only the node in the state named with *With_token* or *With_token_req* has the channel token. The token exchange happens when *TE* is from high to low (Fig. 1.5). This event is triggered by the state transition from *With_token_req* to *No_token* in node B, and will cause the state transition from *No_token_req* to *With_token* in node A. This sequence of transitions transfers the token from A to B without any overlap and completes in finite time (less than one clock cycle). Therefore, conservation of token holds.

The state transition from *With_token_req* to *No_token* is caused by *finish=1*. The conditions for setting *finish* guarantee that *finish* will be one in finite time because the packet length is finite. So this state transition will complete in finite time and the token exchange can also complete in finite time.

Based on the above two properties: conservation of token and finite-time token exchange, we can conclude that it is conflict-free and deadlock-free. □

We can hide the time for token exchange without increasing the latency in the ideal case of no traffic contention. In practice, we design our network router to be pipelined; for example, in a four-stage pipeline the elements might be an input buffer, header decoder, crossbar switch, and output channel. The output channel and the token-exchange are the last stage of pipeline. We can always know if we need to request the channel token before the last stage, for example, in the header decoder stage. Then we may request the token in advance, for example, in the crossbar switch stage. Without traffic contention, we can get the token just in time for the next cycle, *i.e.*, the output channel stage, and hide the token-exchange latency.

We will show in simulation that in spite of the increased overhead of token exchange, the single bi-directional channel has better overall channel utilization, and thus better latency-throughput performance.

1.2.3.2. Link Model Implementation

A physical channel is modeled as a FIFO where the sender pushes a new flit into the channel and the receiver pops a flit out of the channel. There is a delay associated with each channel. When the sender pushes a flit, it will add the delay to the flit's time stamp. The FIFO not-empty signal will inform the receiver to check the channel. If the flit at the head of the channel has a time stamp smaller or equal to the current time, the receiver pops out this flit and completes the transaction. We do not relay an acknowledgment back to the sender because we have the credit scheme described in the previous section which guarantees the buffers will not overflow and no flits will be discarded. The FIFO model of the channel is useful especially when the propagation delay is longer than the cycle time, where there may be more than one flit pipelining on the channel as a result of the transmission line effect. Each flit has its own virtual channel ID. When it is popped out by the receiving node, it will be put into the input virtual channel specified by its virtual channel ID.

The credit feedback link from a receiver back to a sender is similar to the data channel. A credit with specified virtual channel ID is pushed into the credit feedback link. After the scheduled delay, the credit is popped out to update the output credit of the corresponding virtual channel at the sender.

The input buffers are organized as several independent lanes, *i.e.*, virtual channels. The size of buffers and number of virtual channels are the simulation parameters. The output buffer is a single FIFO because the output flits are sequentialized on the physical channel. If the physical channel is grouped as two uni-directional channels, then a single entry latch is sufficient for the output buffer. However, if the physical channel is a bi-directional channel, then we may need more space for the output buffer because the token-exchange may take some time and it is necessary to have the packets pass through the crossbar switch and be buffered at the output port in order to hide the latency.

1.2.4. Traffic Model

A host is connected to one of the ports in a router to inject traffic into the network. The traffic injection module emulates the host to generate packets and encode the packet headers. The injection rate is a parameter to adjust the traffic load in the network. Different injection modes can be used to generate traffic, for example, uniform injection and bursty injection. The packet length is specified in terms of number of flits including the header. Two different packet lengths can be defined and specified by some ratio to generate a mixed traffic. A flit is consumed immediately when it reaches the destination host. The injection rate and destination distribution can be specified for each host separately to model the non-uniform traffic. We may limit the maximal number of outstanding packets per host to prevent overloading the network.

1.3. SIMULATION FLOW

Fig. 1.6 shows the simulation flow of the wormhole data routing network. The simulator has a mixed simulation scheduling scheme. Instead of having the event-driven simulation, we have a global clock to control all the hosts and routers since we assume they are running at the same frequency in the network. We have a bypass in each simulation step to save simulation time if we know there will be no action for some steps. In addition to the global simulation clock, we also have scheduling queues for each physical link as described in the link model. The propagation delay for each transmission is scheduled in the channel queue, and a receiving node will read the channel when the time stamp is reached. The clock skew between nodes is simulated by adding a random synchronization delay when a flit passes across a link. The granularity of propagation delay and synchronization delay can be sub-cycle, but the scheduling is quantized to be on the cycle boundary.

Most network simulations assume it takes only a single cycle for a hop[1,2,5,8,9,15]; this is not an accurate model for hardware implementation and cannot handle of variable propagation delay. In this research, four stages of pipelining in the router are implemented as a realistic model of hardware design. The four stages are shown in Fig. 1.6: I/O Control, Header_Decoding, Request_Arbitration, and Crossbar. The fall-through latency is defines as the latency of a flit passing through a router without contention. In our simulation model, the fall-through latency is four cycles.

In the bi-directional channel configuration, it is very important to hide the delay of token exchange. Since it is known whether an output port will be used at the Request_Arbitration cycle, which is two cycles earlier than the output external transmission, it is possible to issue the token request at the end of the Request_Arbitration cycle if the chosen

```
for (step=0; step < MaxStep; step++)
{
    For (all Hosts)
    {
        Generate_Packet();
        Host_Read_Channel();
    }
    For (All Routers)
    {
        Read_Channel(all input ports);      } I/O Control
        Write_Channel(all output ports);

        Crossbar_Forward();                  } Crossbar

        Requesting(all input ports);         } Request_Arbitration
        Arbitration();

        Header_Decoder(all input ports);     } Header_Decoding
        Write_Input_Buffer(all input ports);
    }
    Cal_statistics(step);
}
```

Figure 1.6: Simulation Flow

output port does not have the token to hide two-cycle latency. Because of the overhead of token exchange, we do not want to have the token exchanged too often but still have to be fair to prevent starvation. A token request is acknowledged when the output port which currently owns the token is idle or is sending a tail flit. It guarantees that a node can get the token in a finite time after it issues the token request, and that the token will not bounce back and forth unnecessarily, thereby reducing the overhead.

1.4. PERFORMANCE MEASUREMENT

Throughput and latency are the main metrics of network performance measurement. Throughput is a measure of the actual data bandwidth delivered to a host. The headers of packets are not included in the net data bandwidth delivered although they will consume the network bandwidth. Assume that the total channel width per port is $2w$. If the channels are bi-directional links, they have entire width $2w$ but may have the token-exchange overhead and longer propagation delay due to higher loading. If the channels are uni-directional links, each direction can only have width w, and the utilization may be lower (Sec. 1.2.3.1).

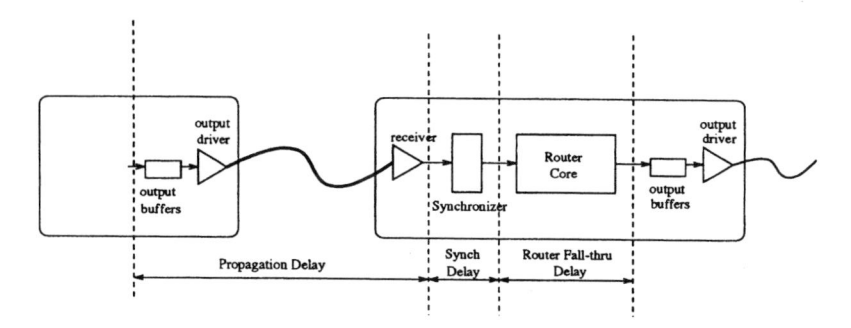

Figure 1.7: Delay components in the simulation model

We normalize the throughput (data bandwidth) to the factor w in our simulation results for comparison of these two channel configuration schemes.

Latency is measured based on each individual flit. Fig. 1.7 shows the delay decomposition of a single hop. The single hop delay consists of propagation delay, synchronization delay, router fall-through delay, and queuing delay (not shown). The queuing delay is due to traffic contention in the network and results in an extra delay which may be dominant at heavily loaded traffic. As described in the previous section, the granularity of propagation delay and synchronization delay can be sub-cycle, and the total delay is accumulated along the path. The router fall-through delay and queuing delay are cycle-based and also accumulated in the latency calculation.

1.5. SIMULATION RESULTS

The simulation was used to study network data routing for two-dimensional tori. Each simulation point has at least 10,000 cycles to ensure that the statistics have reached the steady state. Data for the first 2000 cycles are discarded because they are unstable. For valid comparison, the total buffer size in a node is constant for all simulation conditions, *i.e.* different virtual channel numbers, routing algorithms, and so on. In the simulation, we fix the total input buffer size in a node to be $576w$, where w is the width of a uni-directional channel. For most of the simulations, the packet size is equal to $10w$ ($8w$ data plus $2w$ header). In terms of flits, the packet length is 10 and 5 flits in the uni-directional and bi-directional configuration, respectively. To prevent overloading the network, we limit the maximal outstanding packets per host to be 4 in most simulations (The limitation on the number of outstanding packets exists in most real parallel machines. In the real situation, we usually have request-reply traffic patterns. Then the number of pending requests is limited by the size of request buffer which determines the maximal number of outstanding packets per host.).

Four routing algorithms: Deterministic (dimension-order)[6], Virtual network[12], Dimension Reversal (DRn)[5], , where n is the maximum reversal number allowed, and Star-Channel algorithm[10], have been built in the simulator models. For different routing algorithms, the minimal requirements of virtual channels and buffer allocation to prevent deadlock are different[13]. For example, for the deterministic (Dimension-Order) routing, we need two levels of virtual channels in each port, while the adaptive (Dimension-Reversal DR) algorithm needs three levels of virtual channels. Table 1.1 summarizes how we allocate the buffers for different routing algorithms on a 2-D torus, where vx_i is the number of ith level virtual channels in the x direction. In the table, each virtual channel buffer size (vcb) is $32w$. It is possible to reorganize the buffers to reduce vcb and increase the number of virtual channels. For instance, vx_0=2 with vcb=$32w$ can be reorganized as vx_0=4 with vcb=$16w$. The effect of different vcb size will be shown in the simulation results.

In the simulation, we assume the clock frequency for each node is 100MHz ($10ns$ cycle time) and the propagation delay between nodes is $17ns$. Every node has its own local clock with a different phase, and the clock phase drifts by a random number within ±1% of the clock rate, i.e., $0.1ns$/cycle. The synchronization delay is the clock difference between a sender and a receiver and takes a multiple of $10ns$.

Simulation results and discussion are given as follows:

Effect of routing algorithms We simulated four different routing algorithms to compare their performance under different traffic loads. The four algorithms are listed in Table 1.1: There is one deterministic (Dimension-Order), and three adaptive cases, Virtual Network, Dimension Reversal (DRn)[1], and Star-Channel algorithm. Three different traffic patterns: uniform random, transpose, and hot-spot traffic, were run for each of these algorithms.

---

[1]DRn, where n is the maximum reversal number allowed

algorithm	buffer allocation	total buffer size
deterministic	$vx_0 = 3$, $vx_1 = 2$, $vy_0 = 2$, $vy_1 = 2$	$2(5+4) \times 32w$
adaptive(virtual net)	$vx_0 = vx_1 = vx_2 = 1$, $vy_0 = vy_1 = vy_2 = 1$ for both vn_0 and vn_1	$2(6+3) \times 32w$
adaptive(DR)	det: $vx_0 = vx_1 = 1$, $vy_0 = vy_1 = 1$ adp: $vx_2 = 3$, $vy_2 = 2$	$2(5+4) \times 32w$
adaptive(star)	$vx_0^* = vx_1^* = 2$, $vy_0^* = 2$, $vy_1^* = 1$, $vy = 2$	$2(4+5) \times 32w$

Table 1.1: Virtual channel buffer allocation for different routing algorithms in a 2-D torus

Uniform Fig. 1.8 shows the simulation results of the uniform random traffic for different algorithms. For both uni- and bi-directional channel configurations, the deterministic algorithm is the worst among the four algorithms simulated. The Star-Channel algorithm gives the lowest latency under the same throughput (bandwidth). However, the difference between different algorithms is not very significant compared with some results reported by other researchers[2,5,9]. The main reason for this difference is that we have a more realistic router model where the requests are arbitrated in the same way for both deterministic and adaptive routing so that the hardware complexity is about the same. A second reason is that we have four pipelining stages plus propagation delay in our model. When a flit is blocked and loses 1 cycle due to contention, the percentage of latency increased is smaller than that of the 1-cycle model for a single hop. So the performance difference between algorithms for uniform traffic is not prominent.

Transpose Transpose is a particular data pattern where node (i, j) is always sending messages to node (j, i), and vice versa. Fig. 1.9 shows the results for the transpose data pattern. In Fig. 1.9(a) (Uni-directional channels), the deterministic algorithm saturates much faster than other algorithms due to lack of flexibility in its routing paths. The Star-Channel algorithm is worse than the other adaptive algorithms in this case. The reason is the limitation of usage of non-star channels *i.e.*, the non-star channel buffer must be empty before it can accept any new packet[10]. The efficiency of the adaptive paths drops when we require more adaptivity, which is not very necessary in the uniform case.

For the bi-directional channel case (Fig. 1.9(b)), the deterministic algorithm actually performs best over much of the range of traffic simulated. Because of the regular data flow and the property of transpose, the deterministic algorithm always sends a packet from (i,j) to (j,j) to (j,i). So there is no channel token exchange involved in the path. The only time we may need to exchange tokens is at the source or destination where the router interfaces with the host. For adaptive routing, the regular data flow pattern is destroyed and token exchange is necessary. The overhead of token exchange makes the latency of the adaptive routing algorithms higher than that of the deterministic routing. However, since the network contention is more severe for the deterministic routing, the latency increases rapidly and the network is saturated much more quickly than the case of adaptive algorithms. So the adaptive routing achieves its advantage for heavily loaded traffic.

Hot Spot We also simulated the hot-spot effect in a network. To create a hot spot, we have every host in the network send 2% of its packets to a common destination, *i.e.*, the hot spot. In a 2-D 16 × 16 torus, there normally are about 0.4% of packets for each destination under uniform traffic. Thus we have five times the traffic load for the hot spot. Fig. 1.10 shows the results of hot-spot traffic. For the uni-directional case, all four algorithms saturate at about the same traffic load, which is also different from some results reported previously[2,8]. The first reason is that the bottleneck in this case is at the interface between the router and the host of the hot spot. The limited interface bandwidth will make the contention propagate quickly from the hot spot to other points in the network. This phenomenon is called

WORMHOLE DATA ROUTING IN MULTIPROCESSORS 17

"tree-saturation." Before the tree-saturation occurs, however, the deterministic algorithm is worse than other adaptive algorithms. But as is the case for uniform traffic, the difference is not very significant. When the tree-saturation sets in, most buffers are occupied by the flits destined to the hot spot, so use of different routing algorithms do not make much difference. Additionally, we have limited the number of maximal outstanding packets per host to create a more realistic host model. Since there is more contention for hot-spot messages, their "life time" is longer than normal messages. As a result, based on the packet generation probability, the percentage of hot-spot packets will become much higher than the 2% expected at the steady state. Then there are fewer normal packets which can take advantage of the adaptive routing paths. Therefore, the overall average latency is dominated by the hot-spot packets.

Effect of channel configuration Communication links can be configured either as two uni-directional channels or as a single bi-directional channel (Sec. 1.2.3.1). Fig. 1.11 and 1.12 show the comparison of uni- and bi-directional channels for uniform and hot-spot traffic, respectively. For lower traffic, the bi-directional channel configuration has higher latency because of token-exchange overhead. In this region, latency is dominated by the distance, and any additional delay by token exchange is significant. But when traffic load is increased, the bi-directional channel configuration performs better than the uni-directional configuration because network contention becomes dominant in the latency. For the bi-directional configuration, the effective packet length in terms of flits is half that of the uni-directional configuration, and the traffic congestion is much less in spite of the overhead of token exchange. Especially for the hot-spot traffic (Fig. 1.12), the bi-directional configuration can support much higher data bandwidth because the bottleneck at the hot-spot host is reduced due to doubling the available channel width. A similar comparison for transpose traffic is shown in Fig. 1.16. We will discuss this figure in detail in the paragraph on the effect of packet interference.

Effect of virtual channels The number of virtual channels can be increased to reduce the effect of "blocked-by-head" when there is traffic contention. However, we have to decrease the buffer size for each virtual channel to keep the total buffer number constant for a valid comparison. Dally has shown that more virtual channels can increase achievable throughput[4]. Our simulation shows similar results (Fig. 1.13), where larger buffer size means fewer virtual channels. However, when the buffer size is too small, *i.e.*, when there are too many virtual channels, the performance will be worse. We included propagation delay ($17ns$) and credit feedback scheme in our simulation. When the buffer size is equal to 4 flits, the round-trip link delay will make the sender stop sending more data flits because the credit is 0 and the credit update is delayed even though there is no contention. The channel will remain idle until the credit is updated. So there is an offset between the latency of vcb=4 and larger vcb at lower traffic. Some relative performance gain occurs for vcb=4 at higher traffic, but this is still not the best choice. Therefore, the virtual channel buffer size has to be large enough to hide the propagation delay of credit feedback. In the simulation, vcb=8 or 16 are the optimal buffer sizes for most traffic patterns and routing algorithms.

Effect of packet length For each packet, there is a constant overhead for header routing. When we increase the packet length, the percentage of header overhead is reduced. However, longer packets are easier to be jammed in a network. Fig. 1.14 (a) shows the network bandwidth for packet length=6, 10, and 20 flits. The network bandwidth includes the headers of packets. For shorter packets, the latency is less at the lower network bandwidth compared with longer packets. However, shorter packets have higher header overhead. For example, the header overhead is 1/3 for length=6, but only 1/10 for length=20. Fig. 1.14 (b) shows the data bandwidth which does not include headers. Due to header overhead, shorter packets have lower effective data bandwidth although they can achieve higher network bandwidth. In the simulation, packet length=10 is a good compromise between traffic contention and header overhead for uni-directional channel configuration.

Effect of packet interference When traffic contention occurs, packets will interfere with each other. If more than one packet requests the same output port at the same time (they will request different virtual channels), only one can be granted use of the crossbar to the output port. Different arbitration schemes have been implemented and simulated. The first is that all the flits from different packets have the same privilege to issue the request and only one can be chosen by the wavefront arbiter. Then on an output physical channel, we allow some flits from other packets to break in a packet which was using the channel. The second scheme attempts to keep a packet together. If a packet is using some output port, then the requests by other packets will be filtered out before going to the wavefront arbiter. So the flits from the same packet will be consecutive without interruption by other packets. However, we still allow a flit to break in a packet if this flit is "older" than some threshold to prevent starvation in the second scheme.

For uniform traffic (Fig. 1.15), the performance is better when we allow a flit to break in another packet on a physical channel for the uni-directional configuration. Because all the flits compete with each other only based on their current age, it is a balanced competition and does not depend on previous arbitration results. If we try to keep a packet together, we may make some flits of other packets wait longer even if they are older. For the bi-directional channel configuration, however, keeping a packet together is slightly better except for the deterministic algorithm.

Fig. 1.16 shows the results for transpose traffic. In this case, the benefit by keeping a packet together for the bi-directional channel configuration is even more obvious for adaptive routing. When we keep a packet together, there is no gap within a packet and the packet boundary is clear, so the overhead of token exchange is reduced. If we interleave different packets on a physical channel, a packet will spread out and idle cycles will be inserted when the packet continues to the next node (Fig. 1.17). The output port may be confused by the idle cycles and give up the token prematurely. As a result the token could be exchanged much more often thereby increasing the overhead of token exchange.

1.6. CONCLUSION

The demand of communication in multiprocessors increases rapidly as the network size becomes larger and processors are more powerful. Ac-

(a) Uni-directional

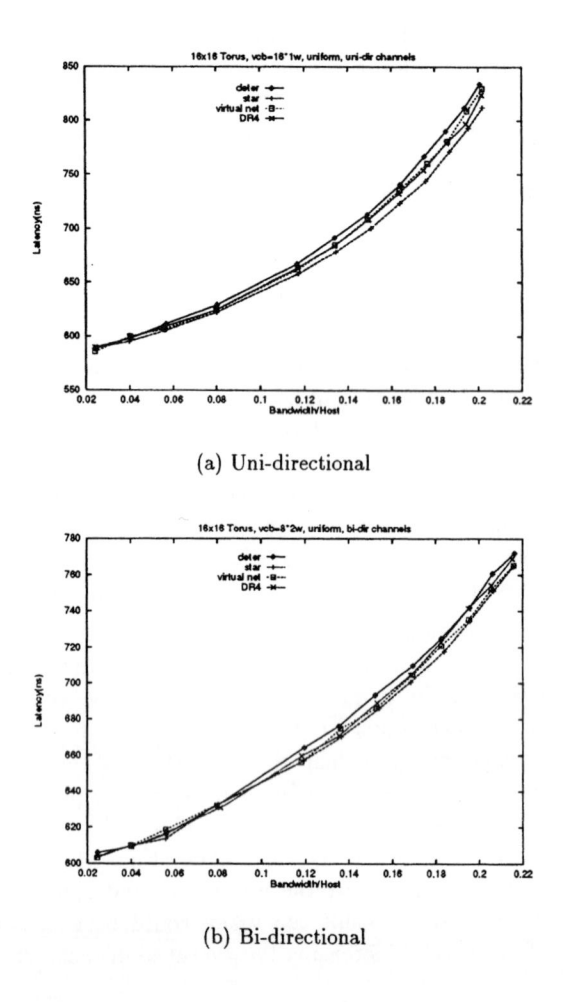

(b) Bi-directional

Figure 1.8: Latency versus Throughput for different routing algorithms under uniform random traffic

WORMHOLE DATA ROUTING IN MULTIPROCESSORS 21

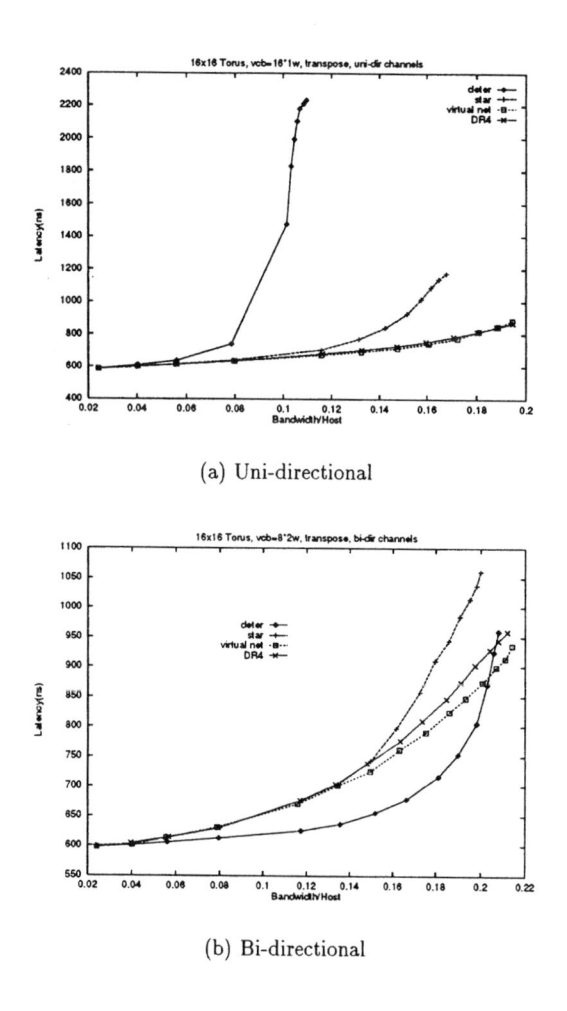

(a) Uni-directional

(b) Bi-directional

Figure 1.9: Latency versus Throughput for different routing algorithms under transpose traffic

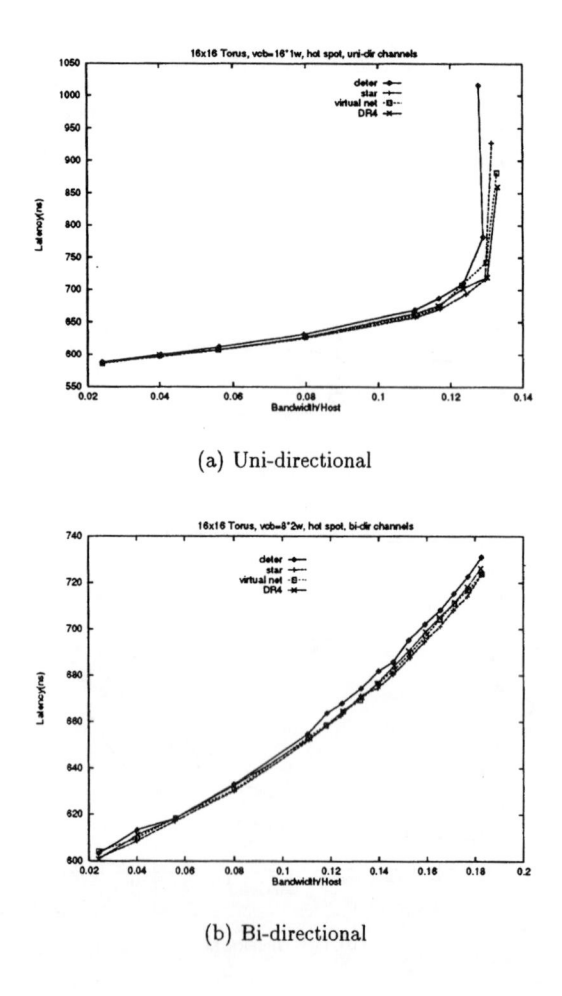

Figure 1.10: Latency versus Throughput for different routing algorithms under Hot-spot traffic

WORMHOLE DATA ROUTING IN MULTIPROCESSORS 23

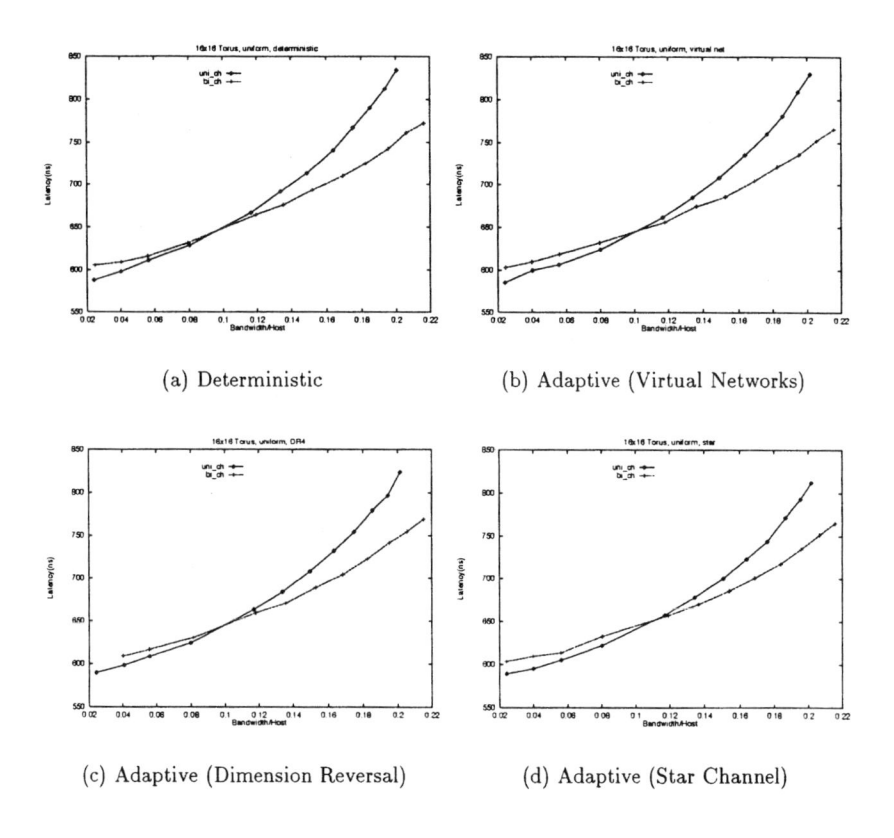

(a) Deterministic

(b) Adaptive (Virtual Networks)

(c) Adaptive (Dimension Reversal)

(d) Adaptive (Star Channel)

Figure 1.11: Latency versus Throughput for different routing algorithms under uniform random traffic. Comparison of uni- and bi-directional channels

(a) Deterministic

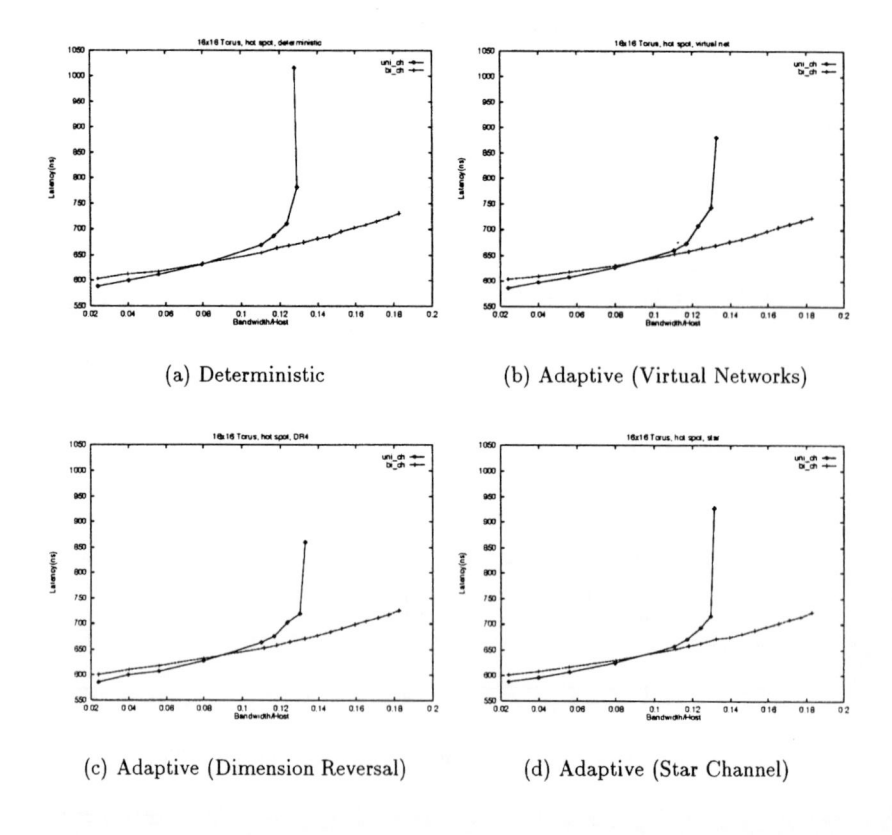

(b) Adaptive (Virtual Networks)

(c) Adaptive (Dimension Reversal)

(d) Adaptive (Star Channel)

Figure 1.12: Latency versus Throughput for different routing algorithms under Hot-spot traffic. Comparison of uni- and bi-directional channels

(a) Uni-directional, size unit=w

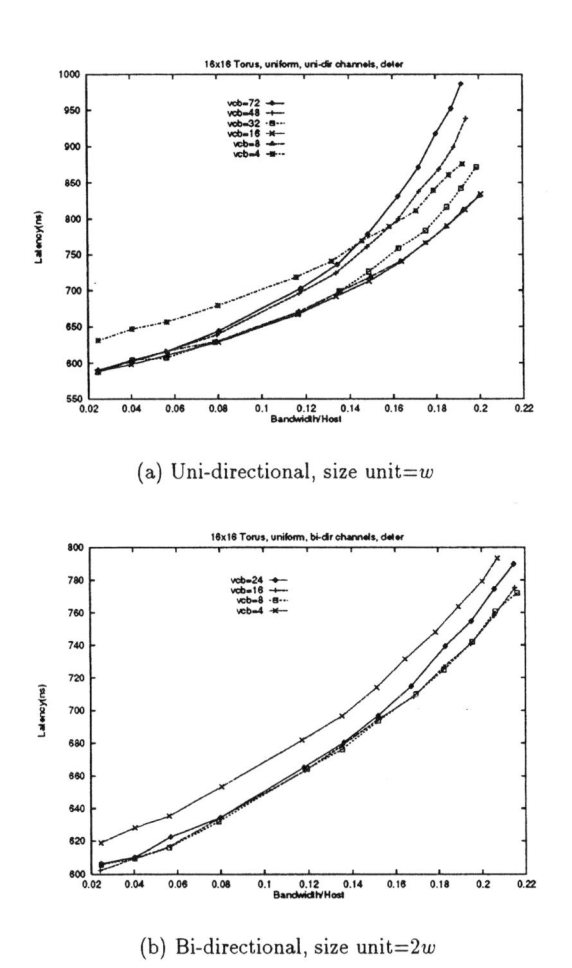

(b) Bi-directional, size unit=$2w$

Figure 1.13: Latency versus Throughput for different virtual channel buffer sizes. Deterministic routing under uniform random traffic

(a) Network bandwidth (including headers)

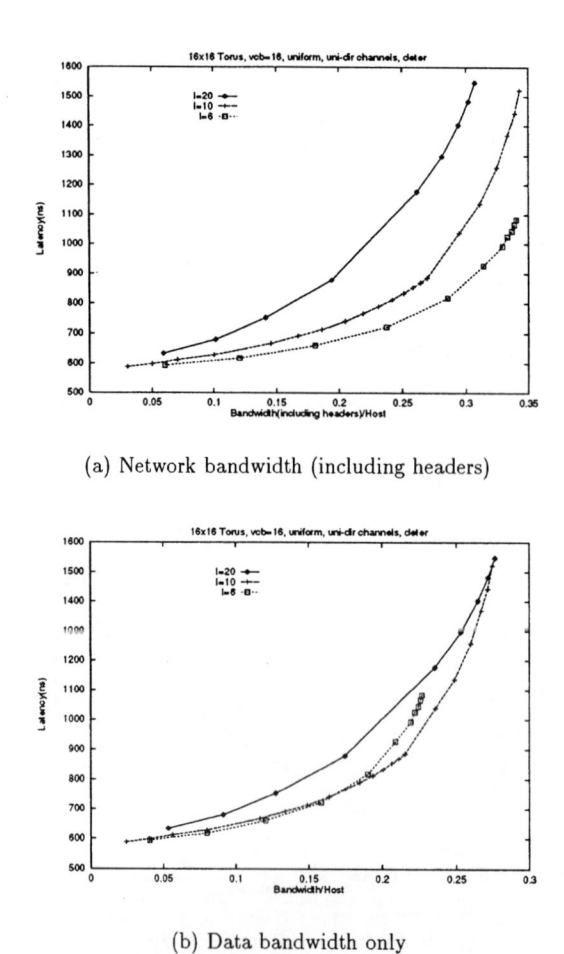

(b) Data bandwidth only

Figure 1.14: Latency for different packet length. Uni-directional channels. Deterministic routing under uniform random traffic.

WORMHOLE DATA ROUTING IN MULTIPROCESSORS 27

(a) Deterministic

(b) Adaptive (Virtual Networks)

(c) Adaptive (Dimension Reversal)

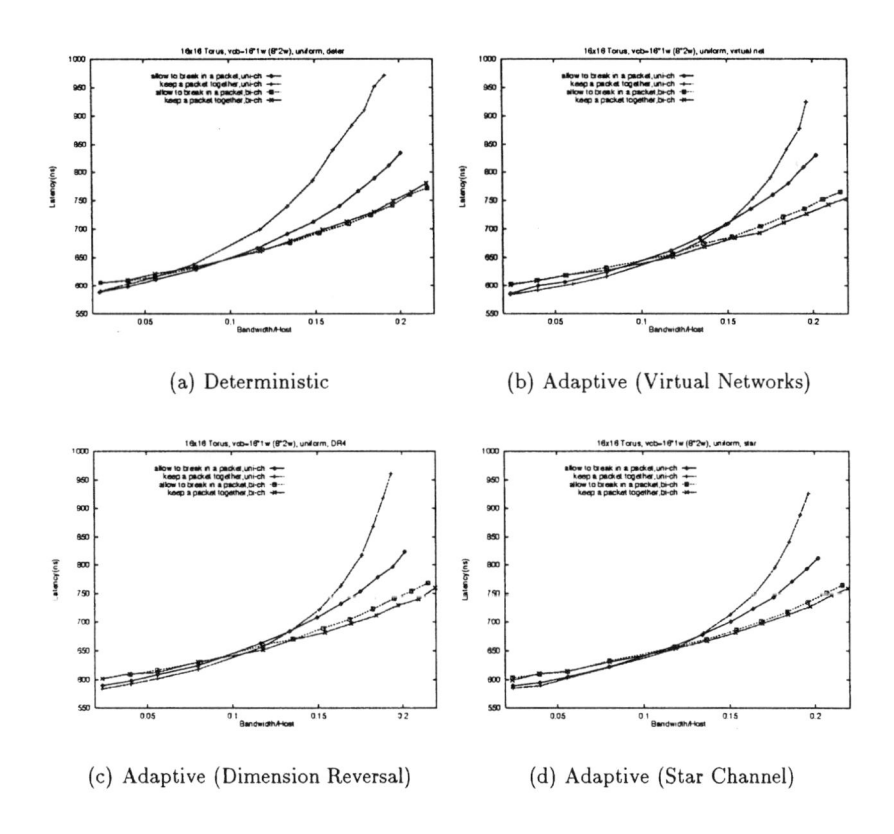

(d) Adaptive (Star Channel)

Figure 1.15: Latency versus Throughput for different routing algorithms under uniform random traffic.

(a) Deterministic

(b) Adaptive (Virtual Networks)

(c) Adaptive (Dimension Reversal)

(d) Adaptive (Star Channel)

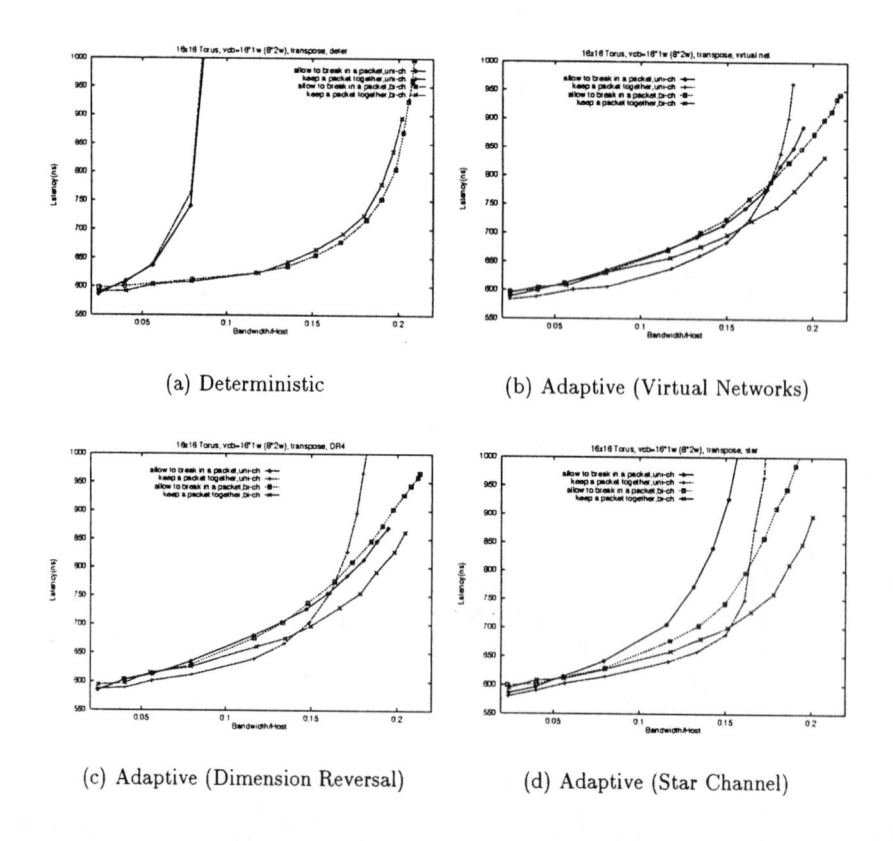

Figure 1.16: Latency versus Throughput for different routing algorithms under transpose traffic.

WORMHOLE DATA ROUTING IN MULTIPROCESSORS

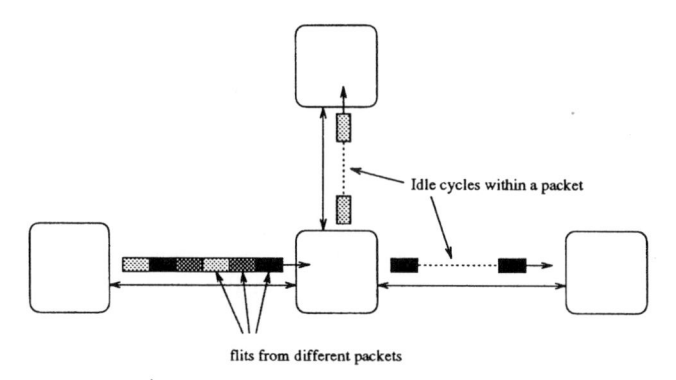

Figure 1.17: Interleaving flits from different packets will insert idle cycles in the packets when they continue to the next node.

curate communication modeling and simulation are the key factors for efficient implementation to achieve high performance. A wormhole routing simulator was built to study different routing algorithms and some design tradeoffs: for example, buffer size, virtual channel numbers, channel arbitration, and so on. We have implemented four stages of pipeline in the router model to simulate a realistic hardware design.

Four algorithms were implemented and simulated: Deterministic (dimension-order), Virtual Network, Dimension Reversal, and Star-Channel algorithm. Three different traffic patterns were included in the simulation: uniform, transpose, and hot-spot. In general, the adaptive algorithms perform better than the deterministic algorithm. The difference between algorithms depends on traffic patterns: for example, it is obvious that adaptive routing is superior for transpose traffic, but not significant for uniform traffic. We also simulated different channel configurations: uni-directional and bi-directional channels. In spite of the overhead of token exchange, bi-directional channels can sustain higher data bandwidth while keeping low latency compared with uni-directional channels. When we use the bi-directional channel configuration, we want to keep a packet together in order to reduce the frequency of token exchange if there is interference between packets. We simulated the tradeoff between number of virtual channels and virtual channel buffer size. In general, more virtual channels can achieve better performance. However, if the virtual channel buffer size is smaller than a threshold (or there are too many virtual channels), the performance degrades because we cannot hide the propagation and credit feedback delay.

REFERENCES

1. P. E. BERMAN, L. GRAVANO, G. D. PIFARRE, and J. L. C. SANZ, "Adaptive Deadlock- and Livelock-free Routing with All Minimal Paths in Torus Networks", in Proc. 4th ACM SPAA, 3-12 (1992).

2. R. V. BOPPANA and S. CHALASANI, "A Comparison of Adaptive Worm-hole Routing Algorithms", in Proc. 20th Int. Symp. on Comput. Arch., 351-360 (1993).

3. W. J. DALLY, "Performance Analysis of K-ary N-cube Interconnection Networks", IEEE Trans. on Comput., C-39(6), 775-785 (June 1990).

4. W. J. DALLY, "Virtual-Channel Flow Control", IEEE Trans. on Parallel and Distributed Systems, 3(2), 194-205 (Mar. 1992).

5. W. J. DALLY and H. AOKI, "Deadlock-free Adaptive Routing in Mul- ticomputer Networks Using Virtual Channels", IEEE Trans. on Parallel and Distributed Systems, 4(4), 466-475 (Apr. 1993).

6. W. J. DALLY and C. L. SEITZ, "Deadlock-free Message Routing on Multiprocessor Interconnection Networks", IEEE Trans. on Comput., C-36(5), 547-553 (May 1987).

7. W. J. DALLY and P. SONG, "Design of a Self-time VLSI Multicomputer Communication Controller", in Proc. Int. Conf. Computer Design, 230-234 (1987).

8. J. T. DRAPER and J. GHOSH, "Multipath E-cube Algorithms (meca) for Adaptive Wormhole Routing and Broadcasting in K-ary N-cubes", in Proceedings of 6th International Parallel Processing Symposium, 407-410 (1992).

9. JOSE DUATO, "A New Theory of Deadlock-free Adaptive Routing in Wormhole Networks", IEEE Trans. on Parallel and Distributed Systems, 4(12), 1320-1331 (Dec. 1993).

10. L. GRAVANO, G. D. PIFARRE, P. E. BERMAN, and J. L. C. SANZ, " Adaptive Deadlock- and Livelock-free Routing with All Minimal Paths in Torus Networks", IEEE Trans. on Parallel and Distributed Systems, 5(12), 1233-1251 (Dec. 1994).

11. K. HWANG, Advanced Computer Architecture (McGraw-Hill, 1993).

12. D. H. LINDER and J. C. HARDEN, "An Adaptive and Fault Tolerant Wormhole Routing Strategy for K-ary N-cubes", IEEE Trans. on Comput., C-40(1), 2-12 (Jan. 1991).

13. Y. W. LU, "Wormhole Data Routing in Multiprocessors: Networks and Algorithms", in State-of-the Art in Performance Modeling and Simulation: Modeling and Simulation of Computer and Communication Networks: Techniques, Tools and Tutorials, edited by G. Zobrist, J. Walrand, and K. Bagchi (Gordon and Breach Publishers, Inc., 1996).

14. L. M. NI and P. K. MCKINLEY, "A Survey of Wormhole Routing Techniques in Direct Networks", IEEE Computer, 26 (2), 62-76 (Feb. 1993).

15. G. D. PIFARRE, L. GRAVANO, S. A. FELPERIN, and J. L. C. SANZ, "Fully Adaptive Minimal Deadlock-free Packet Routing in Hypercubes, Meshes, and Other Networks: Algorithms and Simulations", IEEE Trans. on Parallel and distributed Systems, 5(3), 247-263 (Mar. 1994).

16. Y. TAMIR and H. C. CHI, "Symmetric Crossbar Arbiters for VLSI Communication Switches", IEEE Trans. on Parallel and Distributed Systems, 4(1), 13-27 (1993).

CHAPTER 2

Modeling and Simulation of a Communication Protocol by Stochastic Well-Formed Nets

R. Gaeta, M. Sereno and G. Chiola

2.1. INTRODUCTION

Petri nets (PNs)[1] are an ideal graphical representation of concurrency, synchronization, and resource sharing among asynchronous independent processes. Coloured Petri nets[2] are particularly well suited to the modeling of symmetric systems. Mathematical properties of colored Petri nets can be used for formal model validation purposes as well as for improving the efficiency of performance evaluation[3].

A class of syntactically restricted colored nets called Well-formed nets (WNs)[4] has been proposed. The main analytical motivation for this restriction was the possibility of algorithmically detecting and exploiting model symmetries by means of the concept of *symbolic marking*. Stochastically timed versions of Well-formed nets (SWNs) have been subsequently introduced as well for performance evaluation purposes[5].

From a pragmatic point of view, simulation is the main technique used for the study of large and complex "real" models. Model symmetry properties can be exploited to speedup discrete event simulation as already demonstrated in a prototype implementation[6]. The length of the event list can be substantially reduced in case of large and symmetric models as compared to the usual event-driven simulation technique, yielding substantial improvements in simulation speed in case of large SWN models[7].

This paper describes the application of the SWN formalism to the modeling of a complex ring communication protocol. The protocol modeled is a Medium Access Control (MAC) protocol for Metropolitan Area Network (MAN) called Destination Stripping Dual Ring[8]. The medium is slotted and the topology is dual counter-rotating rings. This protocol has been chosen because it has several characteristics that challenge the efficiency of modeling techniques and tools. It represents then a good test case to assess the actual usefulness of the timed WN formalism. In particular, we are interested in finding examples of construction of models that are at the same time detailed, accurate, fairly easy to understand (compared to the complexity of the protocol behavior), and efficient to analyze and/or simulate.

The paper is organized as follows. Section 2 contains a description of the considered protocol. Section 3 contains a description of the SWN modeling formalism, and of the protocol model. In Section 4 some simulation results are presented and discussed. Section 5 contains conclusions and perspectives of this work.

2.2. THE PROTOCOL DESCRIPTION

The Destination Stripping Dual Ring (DSDR)[8] is a MAC protocol for a network in which the medium is slotted. The traffic is divided into fixed length packets to be inserted into a number of predefined slots that circulate in the rings. Two counter-rotating rings connect the stations. Each station selects one of the two rings to send a packet, choosing the one where the distance to the destination station (measured in terms of number of stations to cross in the rotation sense) is lower. Under the hypothesis of uniform traffic (each station sends packets choosing the destination at random with equal probability) the two rings are used in a completely symmetrical way by each station; we can thus describe the behavior referring to one ring only.

Destination stripping is the key feature for improving bandwidth utilization. The sender inserts its packet into a free slot, and then the packet must occupy the slot until it reaches the destination. The receiver extracts the packet from the slot and strips it, marking the slot as free so that it can be reused to carry other packets. In case of a system connecting N stations, under the above hypothesis of uniform traffic and ring selection, in each rotation period each slot may carry a particular packet while visiting at most $N/2$ stations. Moreover, the slot would carry a particular packet while visiting only $N/4$ stations on the average. The DSDR protocol allows the same slot to be used to carry some other data while passing through the other $3N/4$ stations: the slot can be used two or more times during the same rotation.

The mechanisms set up in DSDR to ensure fairness are: predefined lifetime of the slots (expressed as the number of stations the slot visits before expiring) and a "TOKEN" [1] passing to give stations the right to generate "new" slots to replace the "expired" ones. Stations operate cyclically following the TOKEN rotation. Their behavior is determined by two fundamental events: the arrival and the departure of the TOKEN. During the interval between these two events, the station is said to be "TOKEN-holder" and performs the task of slot generator as described below. Once generated, a slot travels around the ring until it eventually reaches the TOKEN-holder station (possibly different from its generating station). At that moment it "expires," and the TOKEN-holder station "destroys" and replaces it with a "new" slot. Slots having a predefined lifetime can be rewritten as soon as they are stripped but can only carry packets until the end of their (preset) lifetime. A counter contained in the slot header keeps track of the number of remaining stations to be visited by the slot before it expires. The initial value of the counter (denoted K_0) is set when the slot is generated according to the slot lifetime chosen for the network (greater than or equal to the number of connected stations). The computation of K_0 is described in[8]. Each station decrements by one the life counter. A station may write a packet in a slot if the busy bit is 0 and the counter field contains a number $\geq$ than the distance (number of stations) to the packet destination. When a station receives the TOKEN it generates its predefined quota P_{max} of slots before issuing the TOKEN to the next downstream station. While generating slots, if the TOKEN-holder station has packets enqueued for transmission it may insert them in these

---

[1] In order to avoid confusion, in the following we write TOKEN (capital) when we deal with the token of the protocol as opposed to the tokens (lower case) marking a Petri net.

slots. If no data is waiting to be transmitted the slots are generated anyway and left empty. Generating an empty slot means transmitting the proper synchronization bit-pattern; i.e.: a) setting the contents of the counter field to the value K_0 such that if each successive station decreases it by one, the value will reach 0 just before the slot reaches the TOKEN-holder; b) setting the busy field in the packet header to 0 (idle). When a station is not TOKEN-holder it listens to the medium to find slots carrying packets addressed to it, i.e., packets with the busy bit set and the address of that station in the destination field. For such slots, the station must: a) reset the busy bit marking the slot as free; b) decrement the content of the counter field by one; c) read the sender and the data fields from the line. A station that is not TOKEN-holder may send a packet when it finds a slot which is free and whose counter field contains a value higher than the distance from the sender to the destination station. Then it must: a) set the busy bit, thus marking the slot as busy; b) decrease the content of the counter field (which must be greater than 0) by one; c) fill the sender, destination and data fields.

For the correctness of the protocol it is necessary that the expiring slot does not carry an unread packet; this is guaranteed by making sure that the counter field of an expiring slot contains the value 0. Each station is guaranteed the right of sending at least P_{max} slots per cycle. The cycle is the period of rotation of the TOKEN around a ring; it has a constant duration (measured in slots): $T_{cycle} = N \cdot P_{max} + S_{lat}$, where S_{lat} is the round-trip delay (measured in slots). The global behavior of the protocol is completely determined by the following parameters: the physical parameters of the network (i.e., the number N of stations, the latency or round-trip delay S_{lat} and the distance measured in terms of slots between two stations $M = \frac{S_{lat}}{N}$) and $N \cdot P_{max}$, which determines the cycle duration.

2.3. STOCHASTIC WELL-FORMED NETS AND THE PROTOCOL MODEL

We present a rather informal explanation of the WN notation. We refer to[4] for a more complete formal notation.

WNs are a particular case of colored Petri nets as defined in[9]. Some features have been added to the usual colored Petri net structure and a particular syntax is used to specify a Well-formed net: this syntax describes color domains, arc functions and predicates. The introduction of these syntactic constraints allows the application of an effective algorithm for the analysis of the symbolic markings of the model[4]. Despite these syntactic constraints, any colored Petri net model can be re-stated in terms of the Well-formed formalism. A set of basic color classes (i.e. finite sets of colors) indicates the type of tokens present in the places of the net. Each place has a color domain made up of Cartesian products of basic color classes. Transitions are labeled by color domains as well, in order to distinguish different instances of firing for different bindings of the color functions labeling the arcs. Different basic color classes are disjoint and can be *ordered* or not, meaning that a *successor* function (denoted with $\oplus$) is defined or not between pairs of elements of the class. We also define the *predecessor* function (denoted as $\ominus$). Each basic color class is possibly partitioned in *static subclasses*.

The SWN model of the DSDR protocol is depicted in Figure 1. We defined three basic color classes:

- $In = \{n_1, \ldots, n_N\}$ is an ordered class containing one color n_i for each station $(1 \leq i \leq N)$;

- $Tk = \{Token, NoToken\}$ is a non ordered class made up of the two colors used to distinguish the slot carrying the TOKEN;

- Co is an ordered class used to model the set of natural numbers $\{0, \ldots, K_0\}$. It is used to represent both the residual life of a slot and the distance between sender and receiver stations.

Transition **wkload** models the load offered to the system: the projection function $F3 = \langle n \rangle$ binds one component of the transition color domain to one of the corresponding basic colors of tokens in place THINK (a station identity). The meaning is that a station (anyone) "decides" to send a packet. Transitions can be guarded by *standard predicates* constructed as logical expressions concerning equality between projection and/or successor functions, and inclusion in basic subclasses. In the case of transition **wkload**, the output arc determines the distance of the destination station (projection function $F11 = \langle n, d \rangle$); the predicate $[1 \leq d \leq D]$ (where $D = N/2$) restricts the possible values for this distance. The availability of one of the P_{max} slots generated by the TOKEN-holder station is represented by means of a token in place AVAIL_SLOT. In this situation the TOKEN-holder station may have a pending packet to insert in the available slot. Two alternatives are modeled by:

- Transition no_packet that represents the TOKEN-holder station which has no packets to send (the station is "thinking"); when this transition fires its output function $F13 = \langle l, n, t, 0 \rangle$ deposits a token in place RDY_TR with the distance component set to 0 to model an empty slot.

- Transition packet, instead, models the TOKEN-holder station that has a packet to send to a station which is d stations away. The firing of transition packet puts a token in place RDY_TR via function $F12 = \langle l, n, t, d \rangle$; this token represents a full slot which has to visit d station before arriving to its destination.

The deterministic transition pass_next models the synchronous behavior of the protocol. In particular, to model the M steps that a slot has to make in order to arrive to the next station, the service time of transition pass_next is M times the one needed to send a slot.

The successor function $\oplus n$ can be used only in case of ordered basic classes, and binds the projection of the successor basic color of the corresponding components of the transition color domain. Of course the successor function is meaningful only if it is in relation with the projection function n on the corresponding component of the transition color domain. This feature is used by transition pass_next to model the travel of a slot to the next station of the ring.

The transmission of a slot carrying the TOKEN is modeled by the guarded function $F8$ connecting transition pass_next to place SLOT_TKN. Similarly, the guarded function $F7$ model the transmission of a slot that does not carry the TOKEN. The arrival to the next station of a slot carrying the TOKEN is represented by means of a token in place SLOT_TKN while all the other cases are represented by a token in place SLOT_N_TKN.

The state of the stations is recorded in the marking of place TKN_ID, i.e. $M(\text{TKN\_ID}) = \langle n_i \rangle$ means that n_i is the TOKEN-holder station. Upon receiving a slot carrying the TOKEN a station n_k becomes TOKEN-holder; this operation is modeled by the firing of transition change_state that deposits $P_{max} - 1$ tokens in place SCR to model the generation of $P_{max} - 1$ slots that do not carry the TOKEN (through $P_{max} - 1$ firings of transition nc_TOK). Subsequently, n_k generates one

MODELING AND SIMULATION WITH SWNs

slot carrying the TOKEN and this operation is modeled by one firing of transition c_TOK.

When a station receives any type of slot it must:

- Decrease the distance that the slot has to cover to reach its destination. This operation can be done only if the d component of the token representing the slot is greater than zero meaning that the slot is not empty.

- Decrease the residual life of the slot if the l component of the token is greater than zero.

- If the station is TOKEN-holder and the residual life of the slot is equal to zero the slot has to be "destroyed".

The first two operations are modeled by functions $F20$ and $F6$: the first function deals with a slot which has carried the TOKEN while the second function manages all the other cases. $F20$ is a guarded function whose value depends on the value of the component d: when $d = 0$ the slot is empty therefore $F20 = \langle \ominus l, n, d \rangle$ and the decrement of d does not occur, otherwise $F20 = \langle \ominus l, n, \ominus d \rangle$. Function $F6$ behaves in a very similar way for slots that do not carry the TOKEN.

The third operation is modeled by means of transition destroy_slot The destruction can be performed only if the residual life of the slot is null and only if the station is the TOKEN-holder one. These two conditions are enforced by the predicate $[l = 0]$ associated with destroy_slot and by the test arc from place TKN_ID.

A token in place SLOT_NTH models the presence of a slot that does not carry the TOKEN sitting at a station n_k. In this case there may be four possible ways to manage it:

- Transition reuse models the utilization of a free slot by n_k that has a pending packet to send to a destination within a distance less than or equal to the residual life of the slot (the predicate associated with transition reuse imposes this constraint).

- Transition no_reuse represents the case when the destination distance is greater than the residual life of the slot. Also in this case this kind of restriction is modeled by the predicate associated with transition no_reuse.

- Transition full_slot models the forwarding of a slot that carries a packet that still has to cover a distance d to be delivered ($1 \leq d \leq N/2$).

- Transition no_load models the case in which n_k has no packets to send.

As we shall see in the next section, the speed of the simulation process can be improved by modifying the model we presented in the following way: instead of using only place RDY_TR and transition pass_next to model the transmission of slots we separately model the transmission of those slots that carry packets but do not carry the TOKEN by using an additional place (TO_DESTINATION) and an additional deterministic transition (pass_to_destination). Each time a "full" slot is managed by one of the transitions packet, reuse and full_slot a token that represents the slot is put in place TO_DESTINATION; the output arc from transition pass_to_destination to place SLOT_N_TKN is labeled by the function $\langle \ominus^{d-1}l, \oplus^d n, 0\rangle$ to model the arrival of the slot at the destination station. Transition pass_to_destination has a service time equal to $M \cdot d$. In order to preserve the correct behavior of the protocol, place RDY_TR and transition pass_next will still be used to model the transmission of empty slots and of the slot carrying the TOKEN.

2.4. SIMULATION RESULTS

We present some simulation results derived from the model of Figure 1 characterized by the following parameters:

$$N = 20, \quad P_{max} = 3, \quad S_{lat} = 200.$$

The simulation experiments have been performed using the batch method with computation of confidence intervals. The simulations were run until reaching the accuracy of 5% with a confidence level of 99%. The SWN simulator divides a single run in several periods of observation (batches) to gather the statistics about the mean number of tokens in some places and the throughput of certain transitions. The length of a batch is a random variable uniformly distributed between a *minimum length* and a *maximum length*. A transitory period (of the same constant length for each period of observation) is discarded before considering each batch. Both the batch and the transitory length are expressed in terms of number of transitions firings. For our DSDR model we decided to assume the end of the transitory period at completion of the first trip around the ring made by the first slot meaning that all the slots are active around the ring. We estimated the length of this transitory period in 20,000 transitions firings. Furthermore, we have chosen to consider batch lengths randomly varying from 60,000 to 120,000 transition firings that approximately correspond to 3-6 complete rotations of a slot around the ring. The parameter that was varied to obtain different results is the request arrival rate for packets to transmit λ (rate of transition wkload in Figure 1). All timings in the model are normalized to the time needed to send one packet over the ring (arbitrarily assumed to be 1 time unit). The estimated result is the total throughput X observed over the ring which is plotted in Figure 3.

The total throughput is easily identified with the number of firings per time unit of transition wkload in the SWN model in Figure 1 in steady state.

The diagram shows a saturation phenomenon that occurs for normalized throughput 2.83 with this particular choice of N, P_{max}, S_{lat}. The fact that saturation occurs for a throughput value > 2 proves the effectiveness of the protocol in reusing slots within the same round more than twice. The theoretically maximum value of 4 for the normalized throughput in case of uniform traffic appears far from being feasible with the chosen parameters. Larger configurations (with larger values of N, M, S_{lat}) are expected to yield substantially larger normalized saturation throughputs. An extensive study of the protocol characteristics is possible due to the low cost of the simulation results obtained from our proposed model. The simulation experiments have been performed on a Sun SPARC 10/40 equipped with 32 Mbyte RAM and running the SunOS 4.1.3 operating system; the mean CPU time to obtain one point of the graphic was 2109 seconds (about 35 minutes) and the mean number of events considered during one simulation run was 583600 achieving a mean simulation speed of about 300 transition firings per second.

The same set of experiments has been conducted using the modified version of the DSDR model described in Sec.3 obtaining a simulation speed of about 1350 transition firings per second. This result is due to the lower number of events that are scheduled during the simulation runs as a consequence of the use of a deterministic transition with a rate which is dependent on the tokens of its input place.

2.5. CONCLUSION

We presented an example of use of the SWN formalism for the definition and study of a simulation model of the DSDR protocol. The goal of our work was to show the applicability and convenience of the formalism to this application domain rather than studying the protocol itself. In general, a precise definition of a protocol is difficult to transform into a performance model (either analytic or simulative). In this sense we believe that the DSDR protocol is a good example of the entire class of slotted ring protocols. Other protocols of this class should yield SWN models of comparable graphic complexity and comparable simulation cost. The WN formalism not only allows a compact and precise graphic representation of the behavior of the protocol in all its details, but also exploits the symmetry of stations to obtain a high degree of model parametrization. For instance a change in the values of the parameters N, M, and S_{lat} to account for different ring configurations does not imply any change in the graphic model representation. In other words, the graphic complexity of the model is related only to the complexity of the protocol, not to the size of its instantiation on an actual system.

An important advantage of a Petri net based formalism is the possibility of applying different kinds of analysis algorithms to prove qualitative properties (boundedness, safeness, liveness, fairness, etc) as well as derive quantitative results. In this paper we did not provide examples of this kind of analysis, but the interested reader may find in[3] a complete example of combined use of qualitative and quantitative analysis techniques for complex WN models.

From a SWN description of the model a discrete event simulation can be obtained automatically, as implemented in our prototype. The simulation engine is efficient since it is able to capture peculiar characteristics of the model (identified by structural analysis of the underlying WN model). In this respect the level of efficiency is usually comparable to the one obtained by hand coding of a special purpose simulation program by a skilled programmer with a fairly deep knowledge of the system to be simulated. From a practical point of view, fairly accurate performance results may be obtained with a few hours of CPU time on a small and inexpensive workstation on the particular example considered.

From this case study we conclude that the current definition of SWN can be used as a good starting point for the development of a colored

Petri net modeling methodology that can be used efficiently for real applications. The development of some other "realistic" test case could help in identifying extensions of the formalism in order to better match the required modeling power with the availability of efficient analysis algorithms.

REFERENCES

1. T. MURATA, "Petri nets: properties, analysis, and applications", Proceedings of the IEEE, 77(4), pp. 541–580, (April 1989).

2. K. JENSEN and G. ROZENBERG, editors, "High-Level Petri Nets Theory and Application", (Springer Verlag), 1991.

3. G. BALBO, G. CHIOLA, S. C. BRUELL, and P. CHEN, "An example of modeling and evaluation of a concurrent program using colored stochastic Petri nets: Lamport's fast mutual exclusion algorithm", IEEE Transactions on Parallel and Distributed Systems, 3(2), pp. 221–240, (March 1992).

4. G. CHIOLA, C. DUTHEILLET, G. FRANCESCHINIS, and S. HADDAD, "On well-formed colored nets and their symbolic reachability graph", in Proc. 11th International Conference on Application and Theory of Petri Nets, (Paris, France, June 1990).

5. G. CHIOLA, C. DUTHEILLET, G. FRANCESCHINIS, and S. HADDAD, "Stochastic well-formed colored nets and multiprocessor modeling applications", in K. Jensen and G. Rozenberg, editors, High-Level Petri Nets. Theory and Application, (Springer Verlag), 1991.

6. G. CHIOLA, G. FRANCESCHINIS, and R. GAETA, "A symbolic simulation mechanism for well-formed colored Petri nets", in Proc. 25th IEEE-CS Annual Simulation Symposium, (Orlando, Florida, April 1992).

7. G. CHIOLA and R. GAETA, "Efficient simulation of parallel architectures exploiting symmetric well-formed Petri net models", in SCS Western Simulation Multiconference '93, Simulation Series, pp. 285–290, (San Diego, California, January 1993).

8. A. BONDAVALLI, L. STRIGINI, and M. SERENO, "Destination stripping dual ring: a new protocol for MANs", Computer Networks and ISDN Systems, 24(9), pp. 15–32, (March 1992).

9. K. JENSEN, "Coloured Petri nets: A high level language for system design and analysis", in Advances on Petri Nets '90, G. Rozenberg, editor, (LNCS, Springer Verlag, 1991).

MODELING AND SIMULATION WITH SWNs

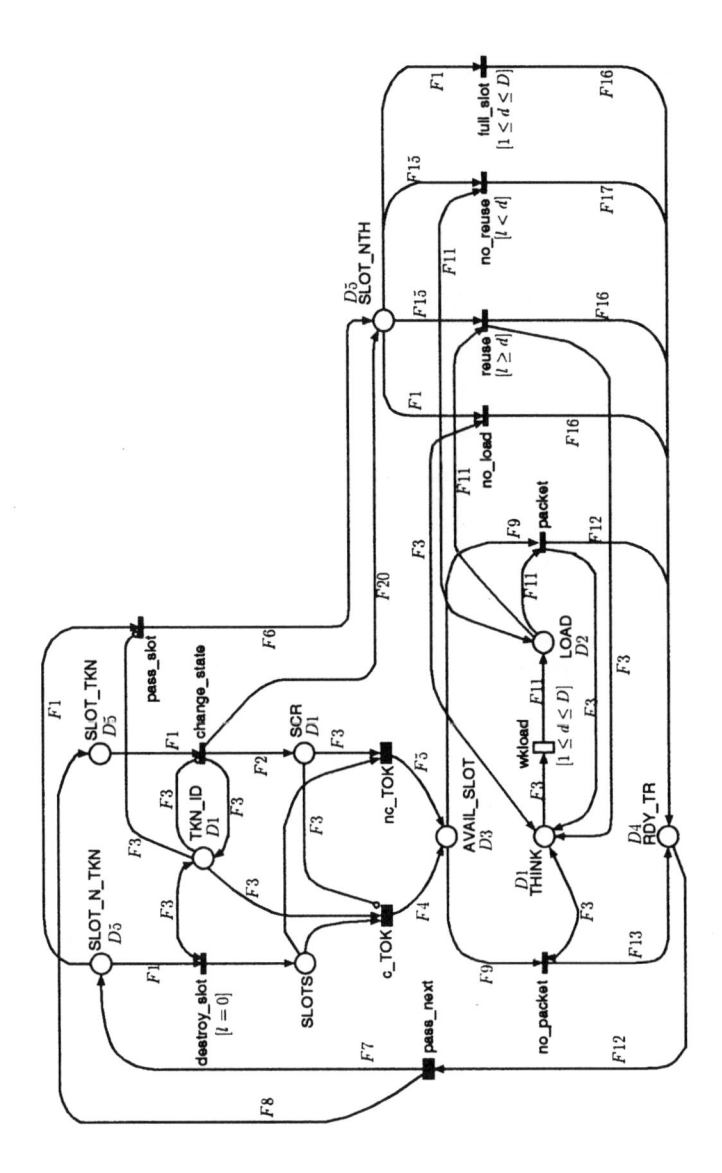

FIGURE 2.1: SWN model of the DSDR protocol.

Function	Definition
$F1$	$\langle l, n, d \rangle$
$F2$	$P_{max} - 1\langle n \rangle$
$F3$	$\langle n \rangle$
$F4$	$\langle K_0, n, Token \rangle$
$F5$	$\langle K_0, n, NoToken \rangle$
$F6$	$[l = 0 \wedge d = 0]\langle l, n, d \rangle + [d \neq 0 \wedge l \neq 0]\langle \ominus l, n, \ominus d \rangle +$
	$+[d = 0 \wedge l \neq 0]\langle \ominus l, n, d \rangle + [l = 0 \wedge d \neq 0]\langle l, n \ominus d \rangle$
$F7$	$[t = NoToken]\langle l, \oplus n, d \rangle$
$F8$	$[t = Token]\langle l, \oplus n, d \rangle$
$F9$	$\langle l, n, t \rangle$
$F10$	$\langle n, S \rangle$
$F11$	$\langle n, d \rangle$
$F12$	$\langle l, n, t, d \rangle$
$F13$	$\langle l, n, t, 0 \rangle$
$F15$	$\langle l, n, 0 \rangle$
$F16$	$\langle l, n, NoToken, d \rangle$
$F17$	$\langle l, n, NoToken, 0 \rangle$
$F20$	$[d = 0]\langle \ominus l, n, d \rangle + [d \neq 0]\langle \ominus l, n, \ominus d \rangle$

Place color domain	Definition	Place	Initial marking
$D1$	In	$M(\text{SLOTS})$	SLAT
$D2$	In, Co	$M(\text{TKN\_ID})$	$\langle n_1 \rangle$
$D3$	Co, In, Tk	$M(\text{SCR})$	$(P_{max} - 1)\langle n_1 \rangle$
$D4$	Co, In, Tk, Co	$M(\text{THINK})$	S
$D5$	Co, In, Co		

FIGURE 2.2: Definitions of the functions, of the place color domanis and of the initial marking for the SWN model of Figure 1.

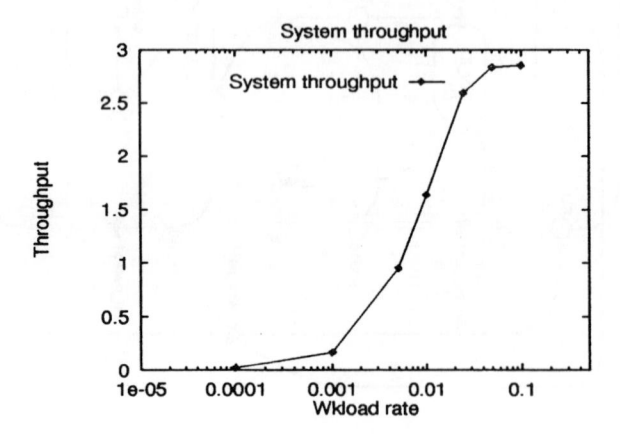

FIGURE 2.3: System throughput vs. rate of transition wkload.

CHAPTER 3

MODELING AND ANALYSIS OF MULTIACCESS MECHANISMS IN SUPERLAN

Adrian Popescu and Rassul Ayani

3.1 INTRODUCTION

Today, new multi-Gbps Local Area Networks (LANs) are designed to support a wide range of applications generating different isochronous and nonisochronous traffic at arbitrary bit rates. The growing demand for high bandwidth networking, under increasing performance constraints, has posed fundamental challenges to LAN design and implementation. In particular, due to the introduction of fiber optic technology, the performance bottleneck is no longer the transmission channel, but rather the network nodes. Three fundamental bottlenecks exist in a multi-Gbbps LAN environment that must be handled in order to achieve optimal performance. These are: the opto-electronic bottleneck, service bottleneck, and processing bottleneck.

The opto-electronic bottleneck results from the fact that networks are inherently limited by the use of electronic components at stations, resulting in performance limitations and inefficient resource utilization. The service bottleneck occurs between the Media Access Control (MAC) layer and the higher layers, and relates to the difficulties in providing the requested

Quality of Service (QoS) for all traffic classes which compete for common transport resources. Finally, the processing bottleneck is caused by the slow (software) processing versus high-speed transmissions.

A novel architectural solution, called SUPERLAN, has been proposed by Popescu[4] to open up all bottlenecks mentioned previously. SUPERLAN is an integrated multi-Gbps LAN where data rates up to 9.6 Gbps are provided in every data channel (isochronous or nonisochronous) for a variable number of stations (up to 60). Total network throughput of about 20 Gbps is achievable. Electronic logic speeds of 100 Mbps and processing speeds up to 20 - 30 MIPS are considered in the design. The total user traffic on the network is separated into two classes, isochronous and nonisochronous, each of which is allocated two or more wavelengths.

A Wavelength Division Multiplexing (WDM) network architecture is considered that is based on the Wavelength-Dedicated-to-Application (WDA) concept. Circuit switching services are considered for isochronous traffic, whereas packet switching services are considered for nonisochronous traffic, which are based on different delay-throughput trade-offs. Switched services at different rates, up to 1.2 Gslot/s, are assumed for each traffic class, for 8 bits/slot or more. Communicative and distributive multi-media services are also taken into account. These may include voice, audio, images, video and data traffic, and may require point-to-point and/or multi-point communications among a variable number of stations with a variable number of substations (at least 10 for every traffic class) connected to each station.

The network model has a physical ring configuration with $(n+1)$ stations. The network has a Master station (S_0) and n Ordinary stations, so-called SUPERLAN stations, denoted by $(S_1 - S_n)$. m substations $\{SS_{i1}, SS_{i2},..., SS_{ij},..., SS_{im}\}$ can be connected to each SUPERLAN station, where SS_{ij} represents the substation j connected to SUPERLAN station i. Each substation generates different types of traffic. Thus, every SUPERLAN station provides a flexible interconnection to different devices, such as multi-media workstations, high-performance computers, high-capacity storage systems, PBXs, diverse audio, image and video devices, with throughputs independent of the network data rates.

The Master station S_0 provides diverse auxiliary functions (clock and frame generation, total loop-length adjustments, resource allocation for isochronous traffic, network management, etc.), while the SUPERLAN stations S_1 - S_n provide communication channels for their (local) traffic. Any SUPERLAN station may transmit and receive simultaneously on both data and control channels.

A single optical fiber (unidirectional link) is used for station-to-station interconnection. A number of eight wavelength channels are used on the fiber.

The network is composed of eight logically separate subnetworks, but provides users with the functionality of a single, integrated multi-Gbps network. It makes use of eight parallel, wavelength-separated channels with time synchronization provided among subnetworks belonging to the same user-traffic class (see Popescu[4]).

A specific solution is proposed for the MAC protocols in SUPERLAN. In this solution, each traffic class/application is provided with its own simple, low-speed, application-oriented MAC protocol, with no interference from other applications. The MAC protocols are separated in the wavelength domain. Their main parameters are chosen based entirely on the application needs of interest. Two control channels, placed at two distinct wavelengths, are dedicated to multiaccess mechanisms for isochronous and nonisochronous control traffic, respectively. Furthermore, two additional channels, placed at two other wavelengths, are dedicated to isochronous and nonisochronous data traffic, respectively.

A Connection-Oriented (CO) procedure with a centralized MAC protocol is provided for the isochronous traffic. The three phases in a CO procedure are supported by different subnetworks in SUPERLAN, i.e., the connection and the termination phases by the control subnetwork, and the data transfer by the isochronous data subnetwork. In the first phase, unknown statistics are considered for isochronous traffic. An admission control mechanism operating at the call level and based on the peak rate for different isochronous traffic classes, both continuous bit-rate (CBR) and variable bit-rate (VBR), is used. The isochronous bandwidth resource available in the data subnetwork (up to 9.6 Gbps) is partitioned into separate bandwidth pools, dedicated to different isochronous traffic classes. This partitioning method is aimed at providing equalization of the blocking probabilities (i.e., fairness) among various traffic classes with different loads and bandwidth requirements. This is based on a multidimensional Erlang loss formula.

3.2 PERFORMANCE MODELING

The network model has a ring configuration (Fig. 3.1) consisting of a master station S_0, n ordinary stations that provide isochronous services $\{S_1, ..., S_i, ..., S_n\}$, and m (isochronous) substations connected to each ordinary station $\{SS_{i1}{}^2, SS_{i2}{}^1, ..., SS_{ij}{}^k, ..., SS_{im}{}^1\}$, where $SS_{ij}{}^k$ represents the substation j connected to station i that provides the subclass k of isochronous service. It is assumed that $SS_{ij}{}^k$ can provide only the isochronous traffic k, which is decided according to different performance experiments. In the case of multimedia substations/terminals, traffic differentiation is still considered according to this model.

In the model, k types of isochronous traffic subclasses, denoted by $\{t_1, ..., t_k\}$, are considered. The k-tuppel $\{b_1, b_2, ..., b_k\}$ denotes the number of temporal slots from the data channel w_d that are allocated (per call) to these subclasses in every 125 µs frame, i.e., b_k corresponds to peak traffic per call for subclass k. For instance, the (CBR) voice traffic needs, in this case, only one slot with a capacity of 64 kbps (i.e., 8 bits/slot), whereas a (VBR) video traffic of type high definition television (HDTV) needs 400 slots of 768 kbps (i.e., 96 bits/slot) every 125 µs frame. This results in a peak traffic of about 300 Mbps.

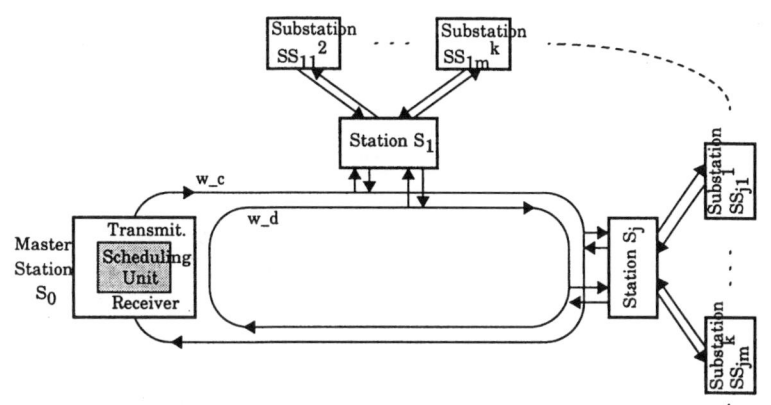

FIGURE 3.1. Network Model for Isochronous Traffic

A specific policy for resource partitioning is used in the master station for allocating bandwidth resource (i.e., temporal slots in w_d) to different classes of traffic and demanding stations. According to this, each traffic class has access to a maximum of $\{w_1, w_2, ..., w_k\}$ temporal slots (of 10 ns each) in every 125 µs frame (bandwidth pools).

The ordinary station S_i is modeled by a multiqueue system with a single cyclic server for the transmission side, and a buffer with two servers for the receive side (Fig. 3.2). A head-of-line (HOL) non-preemptive M/D/1 model with three queues is used to model the transmission side. These queues are dedicated to disengagement requests (priority 2), signaling messages (priority 3), and requests for call setup (priority 4). The highest priority in the transmit multiqueue (Fig 3.2) is given to the incoming upstream traffic, i.e., the incoming control cells from the control channel w_c that are not addressed to that station, and therefore continue further to the next station. Hence, this is a multi-user system with an intermittently available server. In addition, an exhaustive policy is used for serving the accumulated cells in the transmit multiqueue. According to this policy, all newly arrived control cells may be transmitted in the same 125 µs frame, under the condition that they do not exceed the number of cells allowed for that station to be transmitted in one frame.

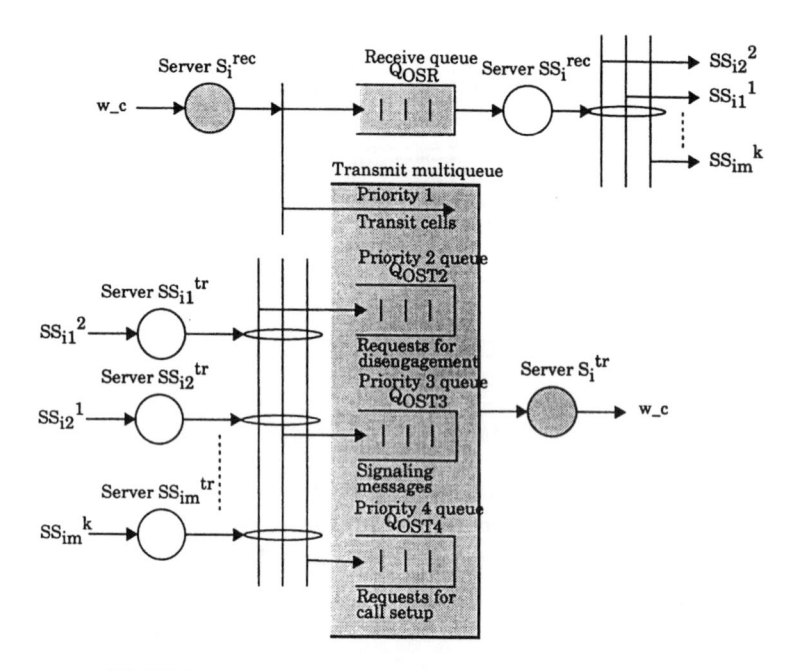

FIGURE 3.2. The Queueing Model for Ordinary Station S_i

The master station S_0 assumes a fork-join model, where the different incoming cells are differentiated, processed (or delayed), and joined for further transmission onto the w_c channel (Fig. 3.3). Four servers and four queues are used to model the master station. The join model has a HOL non-preemptive M/D/1 form, with two queues. These queues are dedicated to signaling messages (priority 1) and MAC messages (priority 2).

A destination removal scheme is used for removing transmitted cells from the w_c channel, i.e., the destination station is responsible for the removal of cells addressed to it.

There are two algorithms for iso MAC protocol, namely in the master station (**procedure** *mac_ms*) and in the ordinary station (**procedure** *mac_os*).

The *mac_ms* algorithm contains five distinct **procedures**: *mac_ms_in* (initialization), *mac_ms_tr* (transmission), mac_ms_rc (reception), mac_ms_set (service of requests for call setup), and mac_ms_dis (service of requests for disengagement). The last four procedures operate concurrently. The mac_os algorithm contains three distinct and specific **procedures** that operate concurrently: mac_os_in (initialization), mac_os_tr (transmission), and mac_os_rc (reception). The interested readers are referred to Popescu[4] for details of these algorithms.

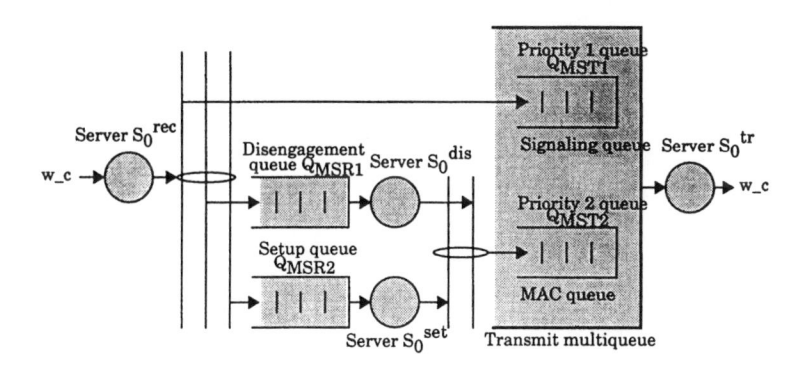

FIGURE 3.3. The Queueing Model for Master Station S_0

A simple complete partitioning policy is chosen for the resource allocation in the w_d channel. The fairness criterion used for the (preliminary) partitioning of resources to different traffic classes (with different bandwidth requirements on w_d) is based on the balancing/equalization of blocking probabilities for these traffic classes, i.e.,

$$\mathbf{P}_{B,1} \approx \mathbf{P}_{B,2} \approx \ldots \approx \mathbf{P}_{B,j} \approx \ldots \approx \mathbf{P}_{B,k} \qquad (3.1)$$

where $\mathbf{P}_{B,j}$ is the probability of blocking for traffic class j.

A similar complete partitioning policy is used for resource allocation, in the w_c channel, among the demanding stations S_i. Accordingly,

$$\mathbf{C}^c = \sum_{i=1}^{n} \mathbf{C}_i^c \qquad (3.2)$$

where C^c is the number of temporal slots in w_c allocated for MAC purposes, and C_i^c is number of cells allocated to station i ($1 \leq i \leq n$). Fairness is provided, in this case, by a specific partitioning of the w_c resource, where C_i^c reflects the percentage traffic intensity demands of station S_i. Accordingly,

$$C_i^c = \zeta_i^c \cdot \Lambda \qquad (3.3)$$

where Λ is the total (average) arrival rate at the master station for call setup requests, and

$$\sum_{i=1}^{n} \zeta_i^c = 1 \qquad (3.4)$$

Also

$$C_i^c \geq \lambda_i \qquad (3.5)$$

3.3 PERFORMANCE METRICS

The performance metrics used are the *call setup delay*, the *blocking probability*, and the *expected number of blocked calls*. The assumptions made in this analysis are:
- Static resource allocation policies are used to assign the w_d and w_c channels to the k traffic classes (data channel allocation), and to the n stations (control channel allocation);
- Exhaustive policies are used for serving the requests for resource in w_c and w_d channels;
- No signaling procedures (and messages) are considered in this model;
- Similar procedures for call setup and disengagement are assumed for all k classes of traffic;
- All substations SS_{ij} are assumed to behave independently;
- The generation of requests for call setup at each substation SS_{ij}^l (providing class l traffic) follows a Poisson process with

SUPERLAN

average arrival rate $\lambda_{ij}^l = \lambda^l$;
- An uniform distribution is assumed in choosing the destination address. According to this, a request for call generated at substation SS_{ij} (where $1 \le i \le n;\ 1 \le j \le m$) is targeted to station S_x (where $1 \le x \le n$; and $i \ne x$) with probability 1/n, and to substation SS_{xy}^l (where $1 \le y \le m$; and $j \ne y$) with probability (1/n) (1/m^l). The parameter m^l represents the number of substations per station dedicated to class l traffic;
- The call holding times for all k traffic classes are assumed to have exponential distribution with average values $(\mu^l)^{-1}$ (where $1 \le l \le k$);
- Calls are considered to be aborted if the substation cannot access the w_d resource in a time period exceeding one second after sending the request for call setup;
- The (transmission and processing) delay between a station and its substations is ignored, since this is a minor part of the total call setup time; and
- All servers (Figures 3.2 and 3.3) have deterministic, constant processing times.

The performance metrics are evaluated as follows.

3.3.1 Setup Delay

The *call setup delay time*, denoted by T for an isochronous call, is the time taken from the instance a substation/user generates a request for call setup to when the substation receives the answer message from the master station. This delay includes:
- the queueing delay W_{QOS} in the queue Q_{OST4} at the ordinary station (Fig. 3.2);
- the access time W_{AOS} at the ordinary station, i.e., the time taken for the first element/cell in queue Q_{OST4} to get its first free (temporal) slot in w_c;
- the transmission times T_{TR} at the ordinary and master stations;
- the transmission delays T_D to and from the master station, including the propagation delay on optical fiber and cross-station delays through the intermediate stations;

- the queuing delay W_{QMSR} in the queue Q_{MSR2} in the master station (Fig. 3.3);
- the processing time T_{MS} at the server S_0^{set} in the master station (Fig. 3.3);
- the waiting time W_{QMST} in the queue Q_{MST2} in the master station (Fig. 3.3); and
- the synchronization latency W_{SYNC} between the data and control channel.

It is assumed that the processing time of server SS_i^{rec} is small enough to avoid queuing in the receive queue Q_{OSR} at the ordinary station (Fig. 3.2). The processing delay of server S_0^{rec} is included in T_D^{iso}. Also, the vacation periods due to the sync header SH, trailer T and gap G fields in the w_c temporal frame are ignored, since their contribution to the call setup delay is negligible. This means that an uniform distribution is assumed in accessing the temporal slots in the w_c channel.

Taking expectations of the above-mentioned times, we have:

$$E[T] = E[W_{QOS}] + E[W_{AOS}] + 2 \cdot T_{TR} + 2 \cdot E[T_D] + E[W_{QMSR}] +$$
$$+ T_{MS} + E[W_{QMSR}] + T_{MS} + E[W_{QMST}] + E[W_{SYNC}] \quad (3.6)$$

where,
- the average waiting time in the queue Q_{OST4} at the ordinary station can be calculated with the formula for a low priority queue in a HOL non-preemptive M/D/1 model (see Bertsekas et al.[2])

$$E[W_{QOS}] = \frac{2\lambda_i (m_{ctr})^2}{2(1-\rho_i)(1-2\rho_i)} = \frac{\lambda_i (m_{ctr})^2}{(1-\lambda_i m_{ctr})(1-2\lambda_i m_{ctr})} \quad (3.7)$$

where m_{ctr} is the (constant) length of the cell (temporal slot in w_c); ρ_i is the traffic intensity at the ordinary station i, which is given by

$$\rho_i = \frac{\lambda_i}{\mu_{ctr}} = \lambda_i m_{ctr} \qquad (3.8)$$

Furthermore, the parameter λ_i represents the average arrival rate for call setup (or disengagement) requests at station i. This is calculated with

$$\lambda_i = \sum_{j=1}^{m} \lambda_{ij} \qquad (3.9)$$

- the average access time $E[W_{AOS}]$ at the ordinary station has two components that are due to the limited capacity available during one temporal frame of 125 μs (fairness considerations) $E[W_{LM}]$, and to the periods of server inability $E[W_{BS}]$. These are busy time slots used by the upstream ordinary stations, to send request cells, and by the master station, to send response cells, with two response cells for each incoming request cell. An exhaustive policy is used in accessing temporal slots in the w_c channel, i.e., every station is allowed to transmit its cells in idle slots as long as the number of transmitted cells in one frame does not exceed a fixed, predetermined number. $E[W_{LM}]$ is calculated with the formula for a M/D/1 model with limited service (see Bertsekas et al.[2]), and $E[W_{BS}]$ is calculated with a formula similar to (3.7).

$$E[W_{AOS}] = E[W_{LM}] + E[W_{BS}] =$$

$$= \left[\frac{n^2 \lambda_i}{2(1-nm_{ctr}\lambda_i)} + \frac{\frac{3}{2}\Lambda + \lambda_i}{2\left(1-\frac{3}{2}m_{ctr}\Lambda\right)\left(1-\frac{3}{2}m_{ctr}\Lambda - m_{ctr}\lambda_i\right)} \right] (m_{ctr})^2 \qquad (3.10)$$

The parameter n is the number of stations in the network, and Λ is the average arrival rate at master station for call setup (or disengagement) requests, which is given by

$$\Lambda = \sum_{i=1}^{n} \lambda_i \qquad (3.11)$$

An average traffic $\Lambda/2$ from upstream ordinary stations is considered to pass through the ordinary station. Accordingly, the average traffic due to the master station is Λ.
- the transmission time for one cell (i.e., the service time for servers S_0^{tr} and S_i^{tr}) is fixed

$$T_{TR} = m_{ctr} \qquad (3.12)$$

- the average value of the transmission delay to/from the master station is modeled as

$$E[T_D] = \left(\left\lceil\frac{n+1}{2}\right\rceil\right)\left(\delta\, m_{ctr} + \frac{3}{2}m_{ctr}\right) = \left(\left\lceil\frac{n+1}{2}\right\rceil\right)\left(\delta + \frac{3}{2}\right)m_{ctr} \qquad (3.13)$$

where the parameter $\delta \cdot m_{ctr}$ captures the propagation delay between stations, and the parameter $(3/2) \cdot m_{ctr}$ is the service time of server S_0^{rec} or S_i^{rec}.
- the processing time in the master station is captured by

$$T_{MS} = \alpha\, m_{ctr} \qquad (3.14)$$

- the average waiting time in the queue Q_{MSR2} at the master station is calculated with the formula for M/D/1 model

$$E[W_{QMSR}] = \frac{\Lambda\,(\alpha)^2(m_{ctr})^2}{2(1-\Lambda\,\alpha\,m_{ctr})} \qquad (3.15)$$

- the average waiting time in the queue Q_{MST2} at the master station approaches zero (i.e., $E[W_{QMST}] \approx 0$) under the assumption that the servers S_0^{set} and S_0^{dis} have equal service times.
- the synchronization delay is averaged over the station position in the ring, with reference to the master station

$$E[W_{SYNC}] = \frac{(\beta + 1)f}{2} \tag{3.16}$$

where the parameter β depends on the number of stations in network and the cell size, and f is the frame size, i.e., $f = 125$ µs. For a number of stations less than 60, $\beta = 1$ (see Popescu[4]).

3.3.2 Blocking Probability

The network is modeled, with respects to the data channel w_d and for class l traffic, as a circuit-switched exchange, where the number of inputs is given by the total number of users of class l, and the number of outputs is given by the maximum number of calls of class l that can be simultaneously serviced. Accordingly, the blocking probability $P_{B,l}^{tc}$ (time congestion) for class l traffic (where $1 \leq l \leq k$), in a generic loss system with X_l inputs (homogeneous sources) and Y_l outputs (where $X_l > Y_l$), can be calculated with the Erlang formula (see Schwartz[5])

$$P_{B,l}^{tc} = \frac{\left(\frac{\lambda^l}{\mu^l}\right)^{Y_l}\binom{X_l}{Y_l}}{\sum_{n_l=0}^{Y_l}\left(\frac{\lambda^l}{\mu^l}\right)^{n_l}\binom{X_l}{n_l}} \tag{3.17}$$

where every user is either idle (in the case of class l traffic) for an (exponential) period of average $(\lambda^l)^{-1}$, or (eventually) generates a call with (exponential) call/session time of average $(\mu^l)^{-1}$. Furthermore,

$$\binom{x}{y} = \frac{x!}{(x-y)!y!} \tag{3.18}$$

is the usual notation for the number of combinations of x objects taken y at a time $(x > y)$.

The parameter X_l is the total number of users in the network that provide class l traffic

$$X_l = \sum_{i=1}^{n} \sum_{j=1}^{m} ss_{ij}^l \qquad (3.19)$$

The parameter Y_l represents the maximum number of calls of class l traffic that can be simultaneously serviced by the network

$$Y_l = \left\lfloor \frac{w_l}{b_l} \right\rfloor \qquad (3.20)$$

The parameter b_l is the peak traffic allocated to call of type l and w_l is the total number of temporal slots in w_d dedicated to traffic class l in every 125 μs frame (bandwidth pool).

As mentioned in section 3.2, fairness is enforced by providing a specific preliminary percentage allotment of the total resource, available in the w_d channel, among the traffic classes. This resource partitioning is done according to the network configuration (i.e., the total number of substations dedicated to different traffic classes) to provide better balancing of blocking probabilities for these traffic classes.

3.3.3 Expected Number of Blocked Calls

The expected number of blocked calls per unit of time (1 hour) for class l traffic is calculated with:

$$\mathbf{N}_{B,l} = n_l^{ut} \cdot \mathbf{P}_{B,l}^{cc} \qquad (3.21)$$

where n_l^{ut} represents the expected number of calls per unit of time for class l traffic, and $\mathbf{P}_{B,l}^{cc}$ is the loss probability (call congestion) of class l traffic. To calculate the parameter n_l^{ut}, we use the formula of traffic intensity for a finite population system (see Körner[3])

$$n_l^{ut} = \rho_s^l \cdot \mu^l = \frac{X_l \cdot \rho_u^l}{1 + \rho_u^l(1 - \mathbf{P}_{B,l}^{cc})} \cdot \mu^l \qquad (3.22)$$

where ρ_s^l represents the traffic intensity per system for class l traffic, and ρ_u^l is the traffic intensity per user for class l traffic ($\rho_u^l = \lambda^l / \mu^l$).

The loss probability $P_{B,l}^{cc}$ (call congestion) for a number of X_l inputs can be calculated with the Engset formula (see Körner[3] and Schwartz[5])

$$P_{B,l}^{cc}(X_l) = P_{B,l}^{tc}(X_l - 1) \tag{3.23}$$

where $P_{B,l}^{tc}(X_l - 1)$ is the blocking probability (time congestion) for $(X_l - 1)$ inputs (eq. 3.17).

3.4 PERFORMANCE EVALUATION

The performance of a class of centralized MAC protocols for isochronous traffic is evaluated in terms of the main parameters of interest: the call setup delay, the blocking probability and the expected number of blocked calls.

3.4.1 Test Conditions

The following test conditions are considered for performance evaluation:
- A cell structure with 74 bits/cell is used in w_c (see Popescu[4]);
- Balanced configuration is assumed for traffic intensity demands from all stations;
- $m_{ctr} = 74 \cdot 10$ ns $= 740$ ns (cell length);
- $C^c = a = 166$ (number of slots in w_c allocated for MAC in one 125 μs frame - equation 3.2);
- $n = 64$ (maximum number of stations);
- $m = 16$ (number of substations per station);
- $C_i^c = 3$ slots/frame for $i = 1$ to 38; and $C_i^c = 2$ slots/frame for $i = 39$ to 64 (resource partitioning in w_c);
- $k = 3$ (number of application/traffic classes);
- $b_1 = 64$ kbps (peak rate for class 1 traffic - telephony

application);
- b_2 = 48 Mbps (peak rate for class 2 traffic - still picture/ graphics application);
- b_3 = 307 Mbps (peak rate for class 3 traffic - application of type color full-screen, full-motion video);
- B_1 = 1 slot/frame with 8 bits/slot (resource allocated to class 1 call in w_d);
- B_2 = 125 slots/frame with 48 bits/slot (resource allocated to class 2 call in w_d);
- B_3 = 400 slots/frame with 96 bits/slot (resource allocated to class 3 call in w_d);
- w_1 = 320 slots/frame (bandwidth pool allocated for class 1 traffic in w_d);
- w_2 = 2900 slots/frame (bandwidth pool allocated for class 2 traffic in w_d);
- w_3 = 9200 slots/frame (bandwidth pool allocated for class 3 traffic in w_d);
- λ^1 = 5 requests/h (average number of requests for class 1 call);
- λ^2 = 3 requests/h (average number of requests for class 2 call);
- λ^3 = 2 requests/h (average number of requests for class 3 call);
- $(\mu^1)^{-1}$ = 3 min (average session time for class 1 call);
- $(\mu^2)^{-1}$ = 20 min (average session time for class 2 call);
- $(\mu^3)^{-1}$ = 30 min (average session time for class 3 call).

3.4.2 Setup Delay

Fig. 3.4 shows the variation of the expected call setup delay time E[T] with the processing time in master station T_{MS}, for different number of stations available on the ring and for a control cell size b = 74 bits/cell. A long distance of about 1.5 km is assumed between stations, which corresponds to δ = 10 (equation 3.13). Also, the service partitioning per station is assumed to be $\{m^1, m^2, m^3\}$ = {14, 1, 1}, where m^l represents the number of substations per station dedicated to class l traffic.

SUPERLAN 63

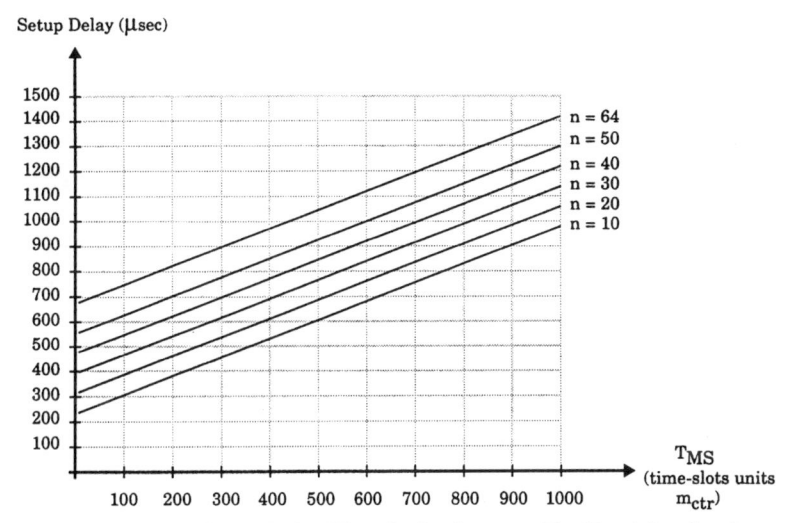

```
Setup Delay (μsec)

1500
1400                                              n = 64
1300                                              n = 50
1200                                              n = 40
1100                                              n = 30
1000                                              n = 20
 900                                              n = 10
 800
 700
 600
 500
 400
 300
 200
 100
      100 200 300 400 500 600 700 800 900 1000   T_MS
                                                 (time-slots units
                                                 m_ctr)
```

FIGURE 3.4. Call Setup Delay Time for Isochronous Traffic with n Stations

Good performance results are obtained for call setup delay times E[T] in that they do not exceed 1.5 ms for even extreme conditions such as assuming long distance between stations, maximum number of stations, high arrival rates for call setup requests, and large processing times in master station to serve the requests for call setup. Also, it is seen from this figure that there is no congestion in the w_c channel. There are therefore no aborted calls because of access delays exceeding 1 second.

These performance results are mainly due to the large capacity available in w_c, whereby the contention for this resource is practically eliminated. The disadvantage, however, is that resource is wasted in the w_c channel. To minimize this, powerful signaling mechanisms acting at a call level and/or during the call can be developed on w_c as well.

The only element that could create congestion in the w_c channel is the processor in the master station S_0^{set} in the case of large processing times and/or sophisticated algorithms for resource allocation (to improve the blocking probability). It is, however, not the case in this model. For instance, about 50 high-level instructions are required to implement the **procedure** mac_ms_set (see Popescu[4]). This corresponds to about 3 Millions of Instructions Per Second (MIPS) in the case $T_{MS} = 100$ m_{ctr}, and less than 1 MIPS when T_{MS} is larger than 300 m_{ctr}. The first congestion limit (i.e., for $n = 64$) is met in the case of very slow processing in the master station, at $3 \cdot 10^{-4}$ MIPS, which corresponds to $T_{MS} \approx 10^6$ m_{ctr}. An average number of four low-level instructions are considered in this case for one high-level instruction. A large reserve of processing capability is therefore available to implement better mechanisms for resource partitioning in w_d and to improve the blocking performance.

3.4.3 Blocking Probability

Fig. 3.5 shows a representative group of analytical curves for the blocking probability $P_{B,3}{}^{tc}$ as a function of offered traffic $\rho^3 = \lambda^3/\mu^3$ for different numbers of class 3 users in the network. The total number of class 3 users in the network is given by $n \cdot m^3$, where n is the number of stations and m^3 is the number of substations per station dedicated to class 3 traffic.

The model is a birth-death process, with arrival rate λ_x^l and departure rate μ_x^l when x calls are in progress:

$$\lambda_x^l = (X_l - x)\lambda^l \qquad for \quad 0 \leq x \leq Y_l \leq X_l \qquad (3.24)$$

$$\mu_x^l = x \cdot \mu^l \qquad for \quad 1 \leq x \leq Y_l \qquad (3.25)$$

where X_l represents the total number of substations in the network that provide class l service, and Y_l represents the maximum number of simultaneous calls supported for class l traffic. When $X_l > Y_l$, the Engset distribution provides the best approximation for the model statistics. However, when $X_l \gg Y_l$ and $\lambda^l \to 0$, the model is better captured with the Erlang distribution (see Schwartz[5]). For $X_l = Y_l$, a binomial distribution is obtained.

Fig. 3.6 shows an example of network dimensioning to obtain fair (i.e., balancing of) blocking probabilities among different traffic classes, in the case of three traffic classes and a fixed resource partitioning in w_d.

SUPERLAN 65

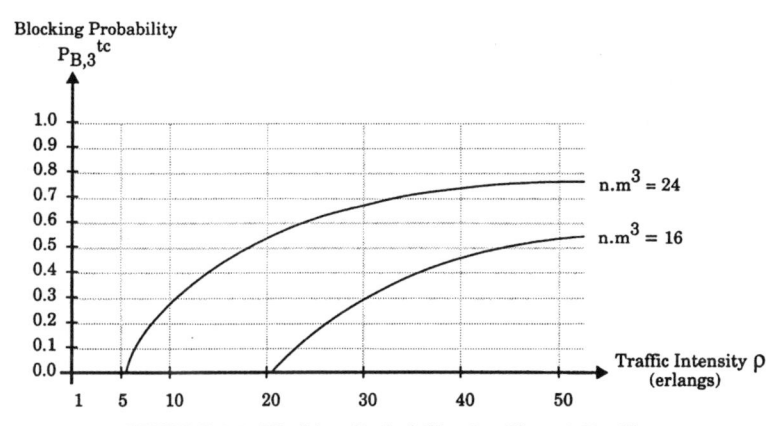

FIGURE 3.5. Blocking Probability for Class 3 Traffic

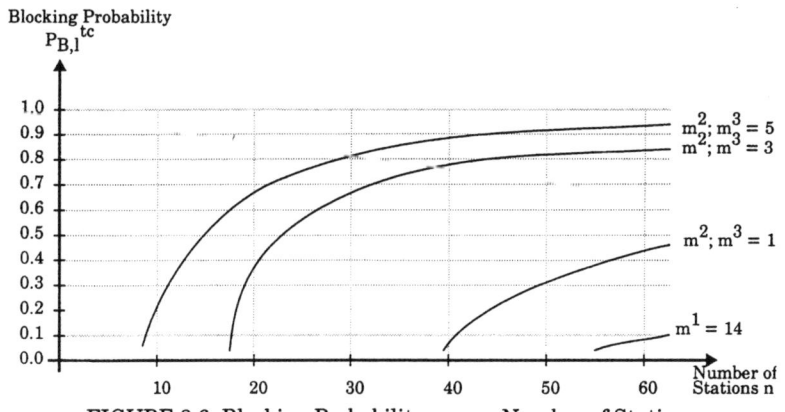

FIGURE 3.6. Blocking Probability versus Number of Stations

The best fairness performance is provided when

$$\{m^1, m^2, m^3\} = \{14, 1, 1\} \qquad (3.26)$$

where m^l represents the number of substations per station dedicated to class l traffic. Also,

$$m = m^1 + m^2 + m^3 = 16 \qquad (3.27)$$

As Fig. 3.6 indicates, the blocking probabilities for all traffic classes are, in this case, the same for a number of stations n up to about 40 and for $\{m^1, m^2, m^3\} = \{14, 1, 1\}$. Beyond this n, $P_{B,2}^{tc}$ and $P_{B,3}^{tc}$ increase more rapidly than $P_{B,1}^{tc}$ because of higher traffic intensity ρ offered for classes 2 and 3.

The main limitation for blocking probability is given by the policy used in the master station for resource allocation that is of the fixed/static resource allocation type. Dynamic access policies, based on dynamic resource sharing mechanisms for resource allocation, must be used to improve the blocking performance. These models provide stations free access to a variable resource in w_d according to instantaneous needs for bandwidth, resource availability, fairness criteria, and access control mechanisms.

3.4.4 Expected Number of Blocked Calls

Fig. 3.7 shows the number of class 3 calls expected to be blocked in an one hour period for different (user) traffic intensities ρ_u and a variable number of users in the network.

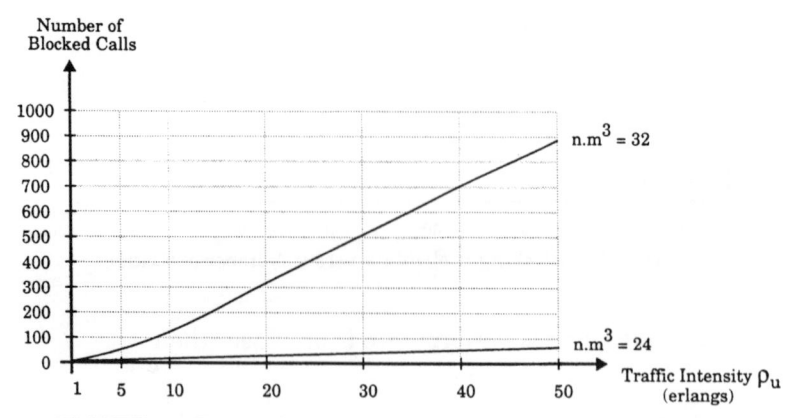

FIGURE 3.7. Expected Number of Blocked Calls for Class 3 Traffic

The parameter $n \cdot m^3$ represents the number of class 3 users in the network, where n is the number of stations and m^3 is the number of substations per station dedicated to class 3 traffic.

3.4.5 Simulation Experiments

A satisfactory performance evaluation of the protocol requires observation of SUPERLAN for at least one hour. Considering the slotted ring structure of SUPERLAN (for instance, with a frame size of 125 µs and slots of 740 ns), a detailed simulation of the protocol would require several weeks of computer time. As an alternative solution, we developed a parallel discrete-event simulator and conducted several simulation experiments. The parallel simulator used both a Conservative Time Window based and an optimistic Time Warp based scheme. Description of the parallel simulator and its performance is beyond the scope of this paper. The interested readers are referred to Ayani et al.[1]

3.5 CONCLUSIONS AND FURTHER RESEARCH

Performance modeling, analysis, and evaluation of a class of MAC protocols for isochronous traffic has been presented. The protocol performance was evaluated in terms of delay for call setup, blocking probability, and expected number of blocked calls.

The performance results show that there is no congestion in the control channel to serve the requests for call setup, and delay requirements are well fulfilled. A large reserve of processing capability is available in the master station to develop specific (and more sophisticated) mechanisms for resource partitioning.

The blocking probability for a simple model, based on static resource partitioning in the master station, has been analyzed and evaluated. The main performance limitation in this case is caused by the policy used for resource allocation, which is a static algorithm and cannot use the resources efficiently. However, we believe that dynamic resource sharing mechanisms can reduce the blocking probability. These mechanisms can provide stations free access to a variable resource, according to their needs for bandwidth, resource availability, fairness criteria, and access control mechanisms. These are topics for further research.

Signaling mechanisms, to complement the MAC protocol, and procedures for multipoint communication, must also be studied and developed. Given the large resource available in the control channel, powerful signaling protocols can be developed to improve system performance.

Finally, another area of interest is to study the separation of the Continuous Bit-Rate (CBR) traffic from the Variable Bit-Rate (VBR) traffic, together with specific media access mechanisms acting at the call and/or burst level. This separation can be done either in the time or in wavelength domain. Similarly, the control channels for these two classes of traffic can be separated in the time or wavelength domain.

REFERENCES

1. AYANI, A., ISMAILOV, Y., LILJENSTAM, M., POPESCU, A., RAJAEI, H. and RÖNNGREN, R.,"Modeling and Simulation of a High Speed LAN," Simulation (Journal of the Society for Computer Simulation), **64(1)**, pp 7 - 14, (January 1995).
2. BERTSEKAS, D. and GALLAGER, R., Data Networks, (Prentice Hall, Englewood Cliffs, NJ 07632, USA, 1987).
3. KÖRNER, U., Tillförlitlighetsteori och Köteori Applicerat på Telekommunikations- och Datorsystem, (Studentlitteratur, Lund, Sweden, 1987).
4. POPESCU, A., A Parallel Approach to Integrated Multi-Gbit/s Communication over Multiwavelength Optical Networks, Ph.D. Dissertation, (TRITA - IT - 9306, Stockholm, Sweden, 1994).
5. SCHWARTZ, M., Telecommunication Networks: Protocols, Modeling and Analysis, (Addison-Wesley Publishing Company, USA, 1987).

CHAPTER 4

MODELING AND MANAGEMENT OF SELF-SIMILAR TRAFFIC FLOWS IN HIGH-SPEED NETWORKS

Ashok Erramilli, Walter Willinger
and Jonathan L. Wang

4.1. INTRODUCTION

In common with findings in many other branches of science and engineering, measurement studies involving working packet networks[1,2,3,4,5] have convincingly demonstrated that the bursty nature of actual network traffic is associated with fluctuations or variations on many time scales; this property is referred to as the *fractal* or *self-similar* nature of traffic[1]. Loosely speaking, the term "fractals" refers to phenomena that vary over many length or time scales. In stark contrast, traditional traffic theory and practice make the implicit assumptions that the fluctuations or variations occur over one, or a limited range of time scales. An example is the Poisson arrival process, in which most of the variation can be said to occur over a single time scale, corresponding to the average inter-arrival time. While the probability of large excursions from the average value is non-zero, the tail of the distribution decays so rapidly that the occurrence of large excursions is highly unusual and does not significantly impact the phenomenon under study. With "heavy-tailed" distributions that underly fractal phenomena, there can be appreciable probability mass many orders of magnitude removed from the average value, and these extreme excursions can dominate system performance even when they occur infrequently.

From experience in other disciplines in science and engineering, phenomena that span many length or time scales present formidable challenges in their description, analysis and control. A popular example is the problem of estimating the length of a coastline[6]: because coastlines exhibit features over a very wide range of length scales, any length measurement depends sensitively on the unit of the yardstick. To this extent, the length of a coastline is arbitrary. A traffic analog of this example is the problem of estimating the "peakedness" (strictly speaking, the asymptotic value of the *Index of Dispersion of Counts* (IDC)) of actual packet traffic; because measured traffic exhibits variability over many time scales of engineering interest, the *peakedness* of a given trace does not rapidly converge to a constant value (as is the case with practically all traditional traffic models) but keeps increasing with the length of the observation interval[1]. To this extent, traditional teletraffic notions such as peakedness are arbitrary, and inapplicable to describe packet traffic. In practice, this means that actual traffic processes can be much more variable than is anticipated by traditional theoretical models. This is just one example of the impact of the fractal nature of traffic (or variations on many time scales) in practice.

One of the objectives of this chapter is to review the state-of-the-art in our understanding of the impacts of realistic traffic on various aspects of *traffic management* for high-speed networks. This assessment is important because current traffic management practice is firmly rooted in traditional assumptions on the nature of traffic, and there exists a considerable gap between "reality" and practice. Experience has shown that often the discrepancies between assumptions and reality are immaterial, and that the traffic management methods are robust. A case in point is the well known *Erlang-B formula* for sizing and analyzing blocking systems such as voice trunks, which is insensitive to the assumptions on the distributions of the holding times. However, traffic management is, in general, an exercise that is only as good as the underlying assumptions on arrival patterns and resource usage. We consider in this chapter several issues in high-speed networking, and assess the impacts of realistic traffic features on them. These issues range from *buffer management* to setting *multiplexing gains*, from *traffic measurements* to *connection admission controls (CAC)*. We present several results that form the basis of newer, and more appropriate traffic management methods for self-similar traffic flows. An exhaustive assessment of these issues is beyond the scope of this report, and indeed the current state of the art. We summarize our understanding of these issues, and indicate on-going work in these areas. It is hoped that this chapter will serve to stimulate further research in this rapidly growing field.

The remaining part of the chapter is structured as follows. In Section 2, we will expand on the fractal nature of traffic, and indicate how the application of self-similar traffic models can lead to parsimonious (that is, low parameter) descriptions of complex bursty traffic phenomena. In Section 3, we consider performance analysis with fractal arrival processes, and discuss two techniques to generate self-similar traffic. In Section 4, we review a number of issues of engineering importance, and assess the impact of discrepancies between reality and practice on these issues. In Section 5, we present topics for future research in the modeling and management of fractal traffic flows in high-speed networks.

4.2. THE FRACTAL NATURE OF TRAFFIC

Recent traffic measurement studies[1,2,3,4,5,7] have identified at least two aspects of actual traffic processes that can be said to span many time scales. One concerns the nature of probability distributions that govern many traffic processes of interest: burst lengths, inter-arrival times, connection holding times, source activity periods or sojourn times, etc. Measurement studies indicate that there is no "typical" or characteristic value that can adequately describes these processes; instead, they can take on values that span many time scales. Mathematically, the variance of these processes is unbounded ("infinite variance syndrome"), that is, the corresponding probability distribution functions are *heavy-tailed*, or using Mandelbrot's terminology[8], exhibit the *Noah Effect*. The second traffic feature that spans many time scales are correlations in traffic processes such as the time series of arrival counts. Correlations, indicating the dependence in traffic activity over time, can span many time scales; mathematically, the autocorrelations decay so slowly that the sum of these coefficients over all lags becomes unbounded. This property is referred to as *long-range dependence* (LRD), or the *Joseph Effect*[8] (in a reference to the Biblical figure who foretold of the "7 fat years and 7 lean years" that ancient Egypt was to experience). LRD implies that there can be considerable *persistence* in bursty traffic processes. Note that queueing backlogs (and perceptions of poor performance) occur in networks when the traffic incident on a resource momentarily exceeds the capacity of the resource. If there were no persistence or correlations in the traffic, it would be improbable for the burst or congestion episode to last for an extended period of time. The backlogs would then be bounded, and the queueing performance acceptable. However, in the presence of persistence, these congestion episodes can indeed be very extended, resulting in considerable backlogs and sharply degraded performance. This intuitive argument demonstrates that network performance can indeed be impacted by fractal characteristics such as LRD. In the next section, we describe several trace driven simulation experiments that support this intuition.

4.2.1. Performance Impacts of LRD

A number of analytical and experimental studies[9,10,11] have established that LRD is a dominant characteristic in determining several performance measures of interest in many network engineering problems. For example, the significance of LRD to network performance is established in [10] by a series of simulation experiments with actual traffic traces, and transformations of traffic traces that preserve some statistical features of the traffic while eliminating others. Readers interested in a detailed description of these experiments are referred to [10]; we present one set of results from these studies to illustrate the performance significance of LRD.

The experiments simulate a queue with the following characteristics: infinite waiting room, deterministic service times, single server, and arrivals taken from a 30 minute Ethernet traffic trace known to possess LRD. The traffic traces consist of time series of counts over 30 millisecond intervals, and in the discrete event simulation, the counts are randomly distributed over the 30 millisecond interval. On an average there are 10 arrivals in each interval. Figure 1 shows the average waiting time in the queue as a function of utilization level, which is varied by changing the capacity of the server. Curve (A) is the delay characteristic obtained with the original trace exhibiting LRD. As can be seen, the delays increase sharply at a utilization level of about 50%. Curve (C) is obtained by completely *shuffling* the traffic trace so that all correlations in the arrival process are eliminated. This is representative of Poisson arrival models that are still widely used in practical traffic engineering. As can be seen, there can be a substantial discrepancy in the load service curve if correlations in the arrival

SELF-SIMILAR TRAFFIC FLOWS

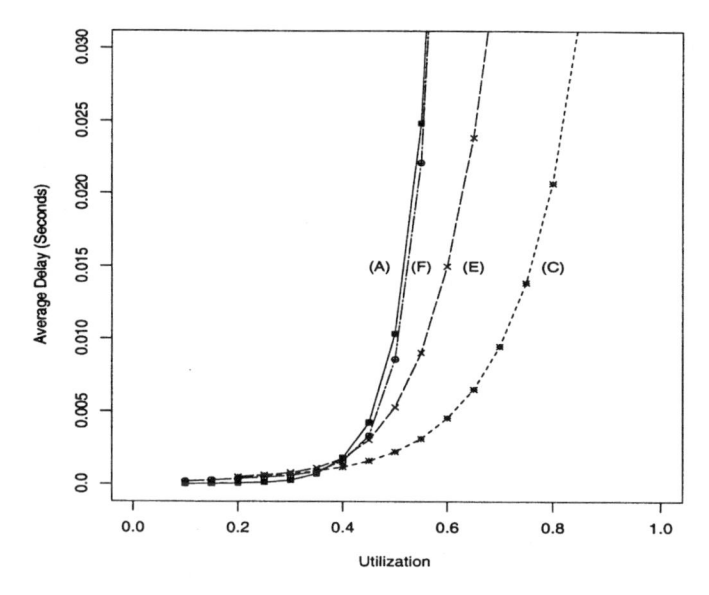

FIGURE 4.1. Queue length distribution: (A) original trace, (C) fully shuffled, (E) external shuffle with a block size of 10, (F) internal shuffle with a block size of 10.

process are not modeled. We now wish to distinguish between the impacts of short-range correlations (which can be effectively modeled by finite-state Markov models) and LRD. We do that by dividing the original traffic trace in blocks of 10, and shuffling the order of the blocks, while preserving the time series within a block. This has the effect of eliminating LRD, while preserving short-term correlations. Curve (E) is the curve that results from using this trace in the simulations. While this curve is closer to Curve (A) than Curve (C), there is nevertheless a substantial discrepancy. We next preserve the order of the blocks while shuffling the order within the blocks. This is representative of a time series that only preserves LRD, while eliminating the short-range correlations. Strikingly, Curve (F) obtained with this trace is almost exactly coincident with the original trace. This experiment makes the point that for many problems of engineering interest (e.g., setting *safe operating points*), LRD is not only relevant, but is a dominant statistical characteristic in the arrival process.

Qualitatively similar results are obtained with: other choices of the block size (from 1 through 50); other Ethernet data sets; working with interarrival traces, as opposed to time series of counts; looking at other performance measures, such as queue lengths, delay percentiles and loss rate; and other (non-Ethernet) traffic traces.

4.2.2. A Self-Similar Model of Data Traffic

One can conclude from the simulation experiments reported in [10] that conventional short-range dependent models which do not incorporate LRD can be significantly in error when used in traffic management. A second conclusion is that the precise structure of short-term correlations is less consequential from the viewpoint of many traffic management problems. Therefore, one can achieve *parsimony* in the model description by abstracting out the actual short-term correlations and modeling the long-term correlations. Note that in this context, the terms "short-term" and "long-term" refer less to the actual time lags, and more to the *structure* of the correlations. In Ethernet traffic traces, for instance, the long-range power-law correlation structure is evident over time scales ranging from a few milliseconds to seconds and minutes, and correlations below this scaling region are referred to as short-term correlations.

There are two ways one can model the long-range correlations that can strongly impact performance. The indirect approach is to model the burstiness at each time scale separately. This is effectively extending traditional modeling approaches to capture the burstiness over all the time scales that are relevant in engineering. This leads, however, often to highly parameterized models requiring a large number of inputs to be specified in practice. The second, more direct approach is based on the assumption that finding fractal or self-similar features in actual traffic traces does not necessarily imply that complicated and highly parameterized traffic models have to be employed to guarantee accurate and relevant solutions to practical engineering problems. In fact, it is well-known that parsimony in models can be achieved by abstracting out features that do not contribute significantly to queueing performance. In this context, it is essential to know what statistical aspects of network traffic can be ignored and when. To this end, we introduce the *Fractional Brownian Motion (FBM) models* for data traffic and provide the conditions under which this model is applicable.

In the FBM model, rather than describing the burstiness on each

SELF-SIMILAR TRAFFIC FLOWS

time scale explicitly, the variation of burstiness over all time scales is modeled parametrically. Specifically, two parameters (a peakedness coefficient a that describes the magnitude of fluctuations on a given time scale, and an exponent called the *Hurst parameter H*) are all that are required. This is the approach effectively adopted by Norros[9] using a Fractional Brownian Motion (FBM) model for the arrival process. FBM can be viewed as an extension of the standard Brownian Motion models that have been used with some success in heavy traffic analysis. In standard Brownian Motion models, the cumulative arrival process $A(t)$ is modeled by random fluctuations about a mean rate m:

$$A(t) = mt + \sqrt{am}Z(t), t \geq 0, \qquad (1)$$

where the process $Z(t)$ has independent Gaussian increments, and a is the term that describes the magnitude of fluctuations. In the FBM model, $Z(t)$ is taken to be Fractional Brownian motion, i.e., the increments of $Z(t)$ are taken to be long-range dependent (and are given by the fractional Gaussian noise). FBM is called an exactly self-similar model, because it has the same burstiness structure on all time scales, i.e., the correlation structure follows the same parametric form over all time scales. In practice, data traffic shows this scaling behavior over a wide range of time scales, though there are natural lower cut-offs (for example, about 10 milliseconds for Ethernet traffic) below which short-range correlations dominate. The FBM model has been shown to be valid under the following conditions:

- the time scales of interest in the queueing processes coincide with the scaling region,
- the traffic is aggregated from a large number of independent users, and
- the effect of flow controls on any one user is negligible.

In particular, the FBM model will in general differ from real traffic in the structure of the short-term correlations, but as has been demonstrated[10], complex short-range correlations can be ignored from a traffic engineering perspective. The long-range correlation phenomenon (or equivalently, the *low frequency structure* of the power spectrum), which is relevant for many aspects of practical traffic engineering, can be parsimoniously modeled using the three parameter FBM model. The next section discusses performance analysis with self-similar processes. While there do exist a limited number of analytical results[9,11,12] which have proved to be very useful in traffic management, techniques to perform routine queueing analysis of self-similar traffic flows are still not available. For this reason, simulation analysis assumes particular importance, and the next section presents several techniques to generate self-similar traffic.

4.3. GENERATING SELF-SIMILAR TRAFFIC

The exact generation of long FBM traces is infeasible in practice due to the amount of storage and CPU time required (consider, for example, the dimensionality of the covariance matrix of the sequence of FBM increments, and the difficulty of generating a sequence that exactly conforms to this correlation structure). Therefore, well understood and efficient approximate algorithms become desirable, especially for generating long traces for the purpose of network performance testing, simulation and analysis. In this section, we briefly describe two such approximate algorithms, namely, the *random midpoint displacement* (RMD) method[13] and an algorithm originally proposed by B. Mandelbrot in an economics setting[14].

4.3.1. The RMD Method

The basic idea in the RMD algorithm is to implicitly build in the dependence structure of FBM by preserving the *scaling* in the displacements of the process around the expected values. Suppose the FBM trace is generated over the time interval $[0, T]$. The RMD algorithm works recursively, subdividing the interval $[0, T]$ and interpolating the values of the process at the midpoints from the values at the endpoints: specifically, the value of the FBM at the midpoint is generated by a random displacement from its expected value, which is the arithmetic mean of its values at the endpoints. The key simplifying assumption in RMD, which results in fast computation at the expense of exactness, is to choose all the midpoint displacements independently. This assumption is obviously valid for ordinary Brownian Motion; for FBM, this is provably not the case. However the RMD construction is reasonable to the extent that the variance of each random displacement (and the overall increment $Z(t) - Z(0)$) scales in the same fashion as FBM.

The variances of the displacements can be calculated as follows. First we let s_k be the standard deviation used in generating the midpoint at step k with σ_0 being the standard deviation of the displacement at time scale T, so that $s_0 = T^{2H}$. We also assume $T = 2^n$. By the scaling properties of FBM, we can show that $s_k = \left(\frac{1}{2^k}\right)^H \sqrt{1 - 2^{2H-2}}\, \sigma_0$ and $s_k = \frac{1}{2^H} s_{k-1}$, with the initial value $s_0 = \sqrt{1 - 2^{2H-2}}$. The computation starts by setting $Z(0) = 0$ and by sampling $Z(T)$ from a Gaussian distribution with mean 0 and variance T^{2H}. Next $Z\left(\frac{T}{2}\right)$ is constructed as the average of $Z(0)$ and $Z(T)$, $\left(\frac{Z(0)+Z(T)}{2}\right)$ plus an offset. The offset is a Gaussian random variable with a standard deviation given by T^{2H} times the initial scaling factor $s_1 = 2^{-H} s_0 = 2^{-H}\sqrt{1 - 2^{2H-2}}$. We then reduce the scaling factor by $\frac{1}{2^H}$, and the two intervals from 0 to $\frac{T}{2}$ and from $\frac{T}{2}$ to T are further subdivided, and the computation proceeds recursively.

The approximate FBM trace generated by the RMD algorithm can be interpreted as the cumulative arrival process $A(t)$ in (1), with m the mean rate, and a the peakedness factor which is defined as the ratio of variance to mean of the number of cells in a unit time interval. The advantages of using the RMD algorithm for self-similar traffic generation are that it is simple, efficient, and fast. Generation of an FBM traffic trace with 260,000 observations only takes about a couple of minutes (including input/output) on a SUN SPARCstation 20. The algorithm also generalizes to a number of other applications: for example, we can measure values of the increments $A(t + 1) - A(t)$ of over coarse time scales and use RMD to compute the process over finer time scales. In a sense, the algorithm can be interpreted as fractal (or self-similar) interpolation. There are, however, also several drawbacks to RMD: the trace cannot be generated "on-the-fly", that is, the whole traffic trace needs to be generated in advance. Secondly, the parameters of the generated sample can be very different from target values. For example, the sample mean rate of the generated trace can be quite different from the target rate. This is because the first end-point generated through RMD determines the sample rate, and this is determined using a single realization of a variable with a standard deviation that can be very large. Note that such variability is inherent in the FBM process. Another potential source of error in the RMD method is associated with the traffic model based on FBM: for $A(t)$ to be interpreted as an arrival process, we need to have *non-negative integer number* of arrivals in a

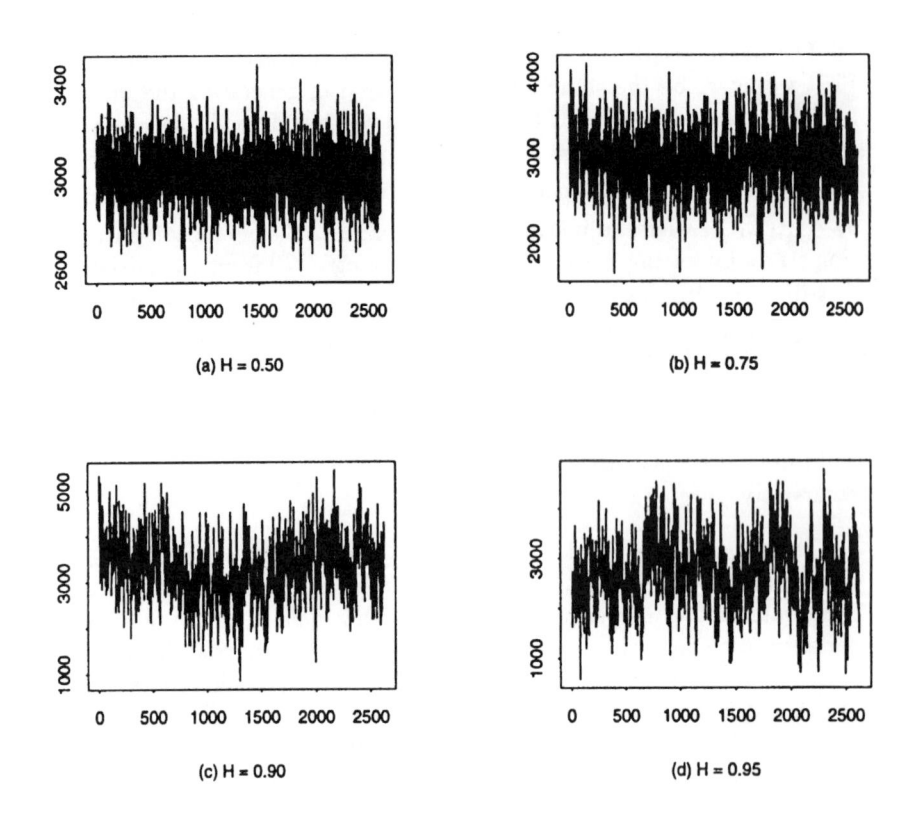

FIGURE 4.2. FBM traces generated using the RMD method: (a) $H = 0.50$, (b) $H = 0.75$, (c) $H = 0.90$, and (d) $H = 0.95$.

unit slot, i.e., the increments of $A(t)$ need to be truncated at zero. This is more generally a limitation of the FBM model, and not specifically of RMD.

As an example, we show in Figure 2 FBM traces with input $H = 0.50$, 0.75, 0.90, and 0.95 ($m = 30$ arrivals per unit interval, $a = 5$ arrivals per unit interval, where unit interval is taken to be 10 milliseconds; note that the unit of a is arrival $\cdot$ (unit interval)$^{1-2H}$). We can see that, as the H value increases, the FBM traces indeed become more and more (long-term) correlated (shown as a low frequency fluctuation in the plot). In contrast, the trace for $H = 0.5$, corresponding to independent increments, does indeed resemble white noise. Thus, the RMD algorithm generates traces that qualitatively resemble FBM. In terms of a quantitative assessment of the RMD algorithm, our analysis[13] indicates that (i) RMD is attractive for qualitative studies and (ii) for quantitative studies the parameters of the generated traces may differ from their target values, due to the natural variability associated with the FBM process. In general, the quality of a given trace can be considerably improved by using aggregated versions of the traces generated by RMD.

4.3.2. Aggregation of many ON/OFF Sources

Developing an approach originally suggested by B. Mandelbrot in an economic context[14], it has been shown recently[7] that the superposition of many *ON/OFF* sources (also known as *packet train models*), each of which exhibits the Noah Effect, results in self-similar *aggregate traffic*. Intuitively, the Noah Effect for an individual *ON/OFF* source model results in highly variable (i.e., infinite variance) *ON*- and *OFF*-periods, i.e., "train lengths" and "intertrain distances" that can be very large with non-negligible probability. In sharp contrast to these findings, traditional traffic modeling, when cast in the framework of *ON/OFF* source models, without exception assumes finite variance distributions for the *ON*- and/or *OFF*-periods (e.g., exponential or geometric distribution). These assumptions result in limited time-scale *ON/OFF* behavior for an individual source, and as a result, the superposition of many such sources behaves like white noise in the sense that the aggregate traffic stream is void of any significant correlations. This behavior is in clear contrast with the measured network traffic described in. Moreover, beyond this mathematical explanation, statistical analyses of Ethernet LAN traffic traces at the level of individual source-destination pairs reported in [7] has demonstrated convincingly that Ethernet LAN data are not only consistent with self-similarity at the level of aggregate packet traffic, but that they are also in full agreement with the physical explanation for self-similarity given above in terms of the nature of the traffic (i.e., the Noah Effect) generated by the individual source-destination pairs that make up the self-similar aggregate packet stream.

An equivalent but deterministic description of this observation in terms of *chaotic maps* is also feasible[15]. Briefly, in the chaotic map formulation, the source state is represented by a continuous variable whose evolution in discrete time is described by a low order, nonlinear dynamical system. The packet generation process is now modeled by stipulating that a source generates a batch of packets at the peak rate when the state variable is above a given threshold, and it is idle otherwise. Realistic *ON/OFF* behavior of single sources can now be described in terms of a small number of parameters associated with a suitably chosen nonlinear map. Either of these methods enables us to model directly what is a plausible physical basis of the self-similarity phenomenon observed in actual network traffic, namely the aggregation of heavy-tailed or highly-variable *ON/OFF* sources.

An immediate benefit for explaining self-similar phenomena in the traffic context in terms of the superposition of many *ON/OFF* sources with infinite variance distribution for the lengths of their *ON/OFF* periods is a straightforward method for generating long traces of self-similar network traffic within *linear* time[7] – assuming a parallel computing environment. Indeed, the above mentioned physical explanation of network traffic self-similarity is well-suited for parallel computing (e.g., every processor of a parallel computer generates traffic according to an *ON/OFF* source with infinite variance *ON/OFF* periods, and simply adding the outputs over all processors produces self-similar traffic) and imitates on the small scale (using a multiprocessor environment) how traffic is generated on a large scale (i.e., in a real-life network).

4.4. ENGINEERING IMPACTS

In the following, we will discuss the impacts of realistic traffic characteristics on a range of significant issues in the traffic management of high-speed networks. These issues include buffer sizing and management, multiplexing gains setting, connection admission controls (CAC), traffic and network controls, as well as engineering considerations such as safe operating points determination, traffic measurement requirements and connection level engineering issues. We are far from a comprehensive resolution of all these issues. As such, the ensuing treatment represents our current understanding of these issues, based on a combination of analytical and simulation studies, as well as physical insights gained from applying self-similar models. The discussion highlights the extent to which assumptions underlying the nature of traffic flows can influence practical traffic management methods.

4.4.1. Buffer Management

In ATM networks, traffic capacity, i.e., utilization levels at which all *Quality-of-Service criteria* (QoS) are met, can be primarily limited by buffer sizes within switching systems. As such, buffer management, the scope of which includes sizing, setting thresholds, measurements as well as discard schemes, is a significant concern in switch design as well as network operations. The importance of buffer sizing was illustrated during early field trials with the first-generation of ATM switches[16], in which cell losses were considerably higher than anticipated. It is now recognized that the limited buffering offered by many of the first-generation ATM switches (as few as 16 cells in some cases) constituted a major design error, and current switches offer considerably more buffering. There are several reasons for the limited buffering offered in many first-generation ATM switching systems. In some cases, smooth traffic flows were assumed because it was anticipated that traffic shaping would eliminated the burstiness of the traffic. In other cases, Poisson assumptions were made for the traffic in the belief that traffic aggregates converged to Poisson traffic. Note that buffers as small as 16 cells can support very low loss rates with Poisson traffic and some vendors still consider Poisson traffic as "bursty" (or even "super bursty"). In practice, the actual traffic incident on the switching systems can be far burstier. If the traffic is exactly self-similar, the bounds on the *queue length distribution*[9,12] can be used in buffer sizing. In such an analysis, we are effectively using the queue length distribution of an infinite buffer system as the surrogate for the variation in cell loss probability with finite buffer size. The cell loss probability with a buffer size of B is approximated by the probability that the queue length exceeds B in an infinite buffer system:

$$P(V > B) \sim \exp[-cB^{2-2H}], \text{ as } B \to \infty, \qquad (2)$$

where

$$c = \frac{(C-m)^{2H}}{2am}\left[\left(\frac{1-H}{H}\right)^H + \left(\frac{H}{1-H}\right)^{1-H}\right]^2. \qquad (3)$$

Here, m, a and H are the parameters of the Fractional Brownian motion model. For short-range dependent traffic, the Hurst parameter H is equal to 0.5, and the queue length distribution decays exponentially, that is,

$$P(V > B) \sim \exp[-cB], \text{ as } B \to \infty. \qquad (4)$$

Thus increases in the size of the buffers cause correspondingly large reductions in the cell loss rate. For most of the traffic traces that we have analyzed in the past several years, H is in the range (0.75, 0.95). For the low end of the range ($H = 0.75$), the cell loss probability varies as

$$P(V > B) \sim \exp[-c\sqrt{B}], \text{ as } B \to \infty. \tag{5}$$

Roughly speaking, keeping all other parameters the same, the buffer requirements are seen to be the square of the prediction for $H = 0.5$. Thus, long-range dependence can have a dramatic impact on the issue of buffer sizing. For example, if traditional models predict a buffer requirement of 50, a Hurst parameter of 0.75 implies that 2500 cell buffers are needed, assuming all other parameters of the decay are unchanged. Thus long-range dependence observed in actual network flows can have a significant impact on buffer sizing.

The next question is whether the "slower than exponential" dependence of the cell loss probability on buffer size, predicted by the FBM model, is observed with real traffic. Figure 3 shows a plot of $P(V > B)$ versus B, using trace-driven simulations, (i.e., curve (A) from Figure 4). Here we have considered a utilization of 0.5, corresponding to the knee of curve (A) in Figure 1. The dashed curve (I) is the form predicted by Equation (2). The values of H and a used to obtain the fit are within the confidence limits of the estimated parameters. The fit between the two curves is very good for almost the entire range shown, and not just the large B regime assumed in theory. At large B, say beyond $B = 100$, the measured queue length distribution falls off faster than the dashed curve, due to limitations of the length of the simulation. As the length of the simulation is increased, agreement between the two curves is stretched out over a greater range. A second issue is the validity of using the queue length distribution of an infinite buffer system as a surrogate for the cell loss probability vs. buffer size function. Figure 3 also presents the results of a *finite buffer simulation* (same Ethernet trace as above, 50% utilization). Clearly, the functional form of the queue length distributions corresponding to the finite (curve (J)) and infinite (curve (A)) buffer systems is the same, with an initial offset, which reflects the well-known fact that the infinite buffer queue length distribution is an upper bound on the finite buffer loss curve. It is interesting to note that the "slower than exponential decay" appears to be preserved in the finite buffer case, though mathematical analysis to support this conjecture are lacking at present.

SELF-SIMILAR TRAFFIC FLOWS

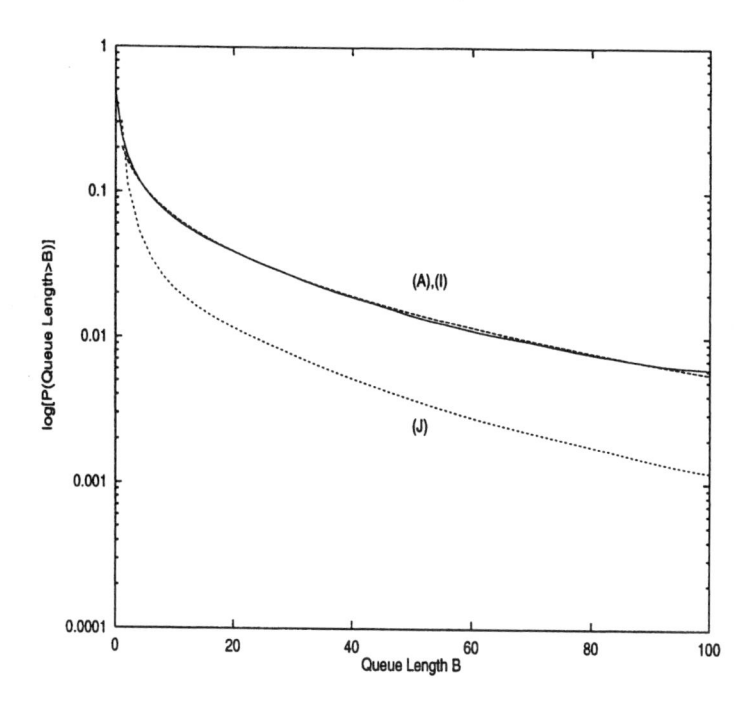

FIGURE 4.3. Queue length distribution: (A) original trace, (I) predicted by Equation (4), (J) finite buffer simulation using original trace.

From Equation (2), we see that with realistic traffic, an increase in the buffer size can have a relatively small effect on the cell loss probability. This is referred to as *buffer ineffectiveness*[17], which might be (incorrectly) interpreted to imply that no amount of buffering will be adequate to achieve the low loss rates that are required in ATM. Note that the plots in Figure 3 are obtained at relatively high utilizations, when the queueing performance is already deteriorating. At these levels, which have already exceeded "safe operating points" (discussed later in Section 4.5), even small increases in load can have large impacts on performance, and buffering is no therefore no longer "effective". The real benefits of buffering should not be assessed when the system is "in trouble" and queueing performance is already unacceptable; but it should be assessed by the improvement in loading levels

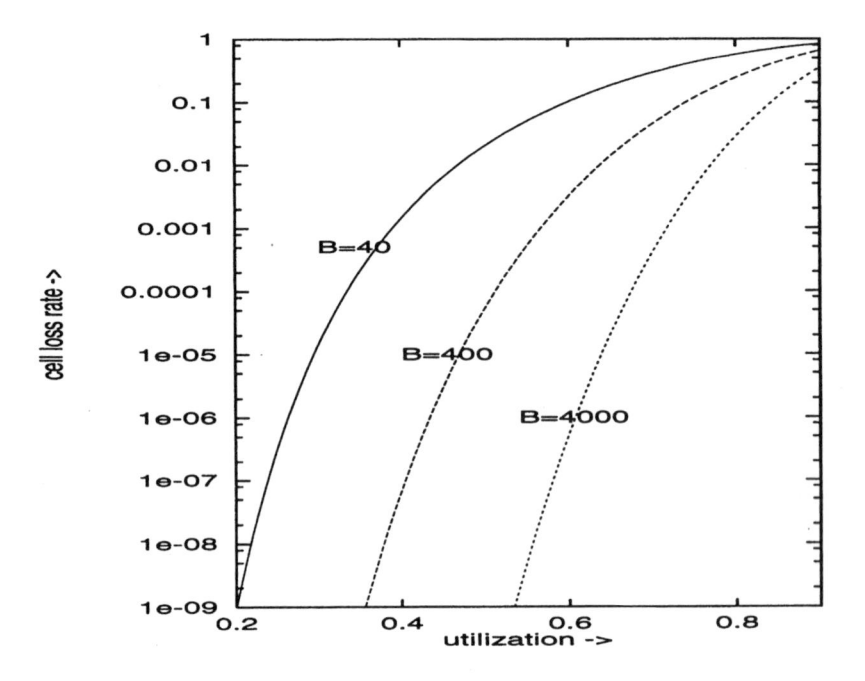

FIGURE 4.4. Cell loss rate vs. utilization levels for buffer sizes of 40, 400, and 4000.

(when the system is "safe" or "stable") that are permitted by increasing buffer sizes. This is shown in Figure 4, which shows the cell loss rate vs. utilization levels for three buffer sizes: 40, 400 and 4000. For a given cell loss-threshold (say 10^{-7}), the operating point increases with buffer size. In particular, increasing the buffer space from 40 to 4000 increases the operating point from about 24% to excess of 57%.

4.4.2. Statistical Multiplexing Gains

A critical issue in broadband traffic management is deciding appropriate levels of multiplexing gains. Some form of statistical multiplexing, or "overselling", is essential for reducing networking costs. The use of packet or cell switching imposes considerable overheads, in the form of bandwidth used to transport cell and packet headers (e.g., in ATM, cell header constitutes 15% overhead), and the processing involved in switching on a per cell or packet basis. This extra overhead can be justified on the basis of more efficient usage of transmission facilities and switching systems through statistical multiplexing. In current broadband networks, overall capacities and costs are typically limited by the availability of switch ports. The use of statistical multiplexing gains in the access and backbone portions of the network also results in more efficient use of switch ports.

The FBM model can, in fact, be used to set safe levels of statistical multiplexing gains. Based on the FBM model, we can show that the Hurst parameter H is preserved under multiplexing of identical sources. In addition, when heterogeneous sources, with identical mean traffic rates and "peakedness" but different H values, are multiplexed, the largest H will dominate. The interpretation of the Hurst parameter as a measure of burstiness, along with this non-decreasing property, have led some to conclude that multiplexing does not reduce the burstiness of traffic, and as such multiplexing gains are not feasible with self-similar traffic. On the contrary, the FBM model does predict significant multiplexing gains when a sufficient number of independent sources are multiplexed. This is because burstiness is characterized not just by the correlations in the fluctuations parameterized by H, but also by their *relative* magnitude given by $\sqrt{a/m}$. When n independent sources are multiplexed, the relative magnitude is reduced by $\sqrt{n}$. Therefore, in environments where the capacity of a server is significantly greater than that of the magnitude of fluctuations of a single source, substantial multiplexing gains can be realized. There are two ways of realizing multiplexing gains over a link supporting a number of independent bursty sources: "within" a source, and "across" sources. In the former, as the traffic fluctuates between peaks and valleys, one attempts to gain efficiencies by allocating a given traffic source capacity less than its peak rate. When the source bursts at levels greater than its allocated capacity, buffering is used to store the cells. In effect, the buffers in an ATM network serve to smooth out the "peaks" in a traffic stream, and the implicit assumption is that the traffic source does not exceed its allocated capacity for extended periods of time. In multiplexing gains "across" sources, one attempts to exploit the fact that the "peaks" in a stream may coincide with the "valleys" of another independent stream. Thus the total capacity required to support n streams will be considerably less than the sum of the capacities required to support each source, resulting in multiplexing gains.

The essential insight gained from an analysis of fractal traffic models is that (i) significant gains are not possible in multiplexing within a source (ii) however, there is considerable potential for gains across sources. In contrast, with traditional models, even limited amounts of buffering can result in considerable gains within a source, so much so that, much of the earlier work[18,19] on the well-known *equivalent bandwidth approach* ignored multiplexing gains across sources. However, the persistence observed in real traffic traces implies that "peaks" and "valleys" do not follow in a rapid succession, and intervals in which a particular source exceeds a given allocation can be very extended. When this happens, very large buffers are required to smooth out fluctuations on such extended time scales. Note that ignoring multiplexing gains across sources will result in very conservative settings of multiplexing gains with self-similar traffic flows.

4.4.3. Connection Admission Controls

An issue closely related to safe operating points and setting statistical multiplexing gains is connection admission control (CAC). The objective of a CAC is to evaluate the bandwidth and QoS requirements of a connection request against available capacity, and to make the decision on whether or not to accept the connection. The call is accepted if the CAC algorithm indicates that doing so will not degrade the QoS of this and other (already existing) connections to unacceptable levels. Typically, a CAC estimates the bandwidth needed by a new connection, and accepts the connection if the residual capacity is sufficient to meet the desired QoS. In a linear equivalent bandwidth scheme, the effective bandwidths are additive, and are computed solely on the basis of traffic descriptors of the connection and system parameters; in particular, the residual bandwidth is a "sufficient statistic" that captures the effects of the already existing connections. While this is a great simplification in practice, such schemes ignore multiplexing gains across sources, and as indicated earlier, will result in very conservative admission controls for self-similar traffic flows. In nonlinear schemes, the effective bandwidth needed to support a connection depend on the characteristics of the other connections. This results in greater complexity in the algorithm, which may however be needed to realize reasonable efficiencies. As

SELF-SIMILAR TRAFFIC FLOWS 87

with other traffic management issues, the self-similar nature of traffic can have a significant impact on the effectiveness of CACs algorithms, and motivates the development of algorithms that achieve efficiencies and are practical.

4.4.4. Traffic Controls

Guaranteeing an often stringent set of Quality-of-Service (QoS) requirements, maintaining service stability and realizing efficiencies requires a variety of controls at the network and application levels, besides CACs. The interactions between traffic flows and controls at the network and application level are significant for traffic management. For example, an important question to address is whether a combination of *shaping* and *policing* can eliminate long-range dependence. However, shapers are ideal "low pass" filters[20], which means that they are most effective in smoothing out high frequency fluctuations (which are relatively inconsequential for performance), but ineffective in eliminating LRD. Specifically, eliminating LRD will require enormous buffering that will introduce significant delays at the shaping mechanism, thereby offsetting any of the benefits of smoother traffic flows within the network. Further results on the effects of shaping and policing in the presence of fractal traffic can be found in [10,20]. For example, given that end-to-end QoS objectives are to be met, it has been shown in [10,20] that with shaping (i) the Hurst parameter, which is related to the asymptotic properties of the traffic stream, is unchanged, and (ii) the overall variability of the traffic can be reduced by eliminating the high-frequency fluctuations, but this does not result in significant improvements in efficiencies. In contrast, traditional traffic models can predict significant efficiency improvements using shaping. Thus, the nature of traffic assumptions can lead to very different prescriptions for efficient networking architectures. Traditional traffic assumptions would lead to a reliance on shaping at the edge of the network, based on "relocating" a modest number of buffers to support this function, that would result in smooth traffic flows and high efficiencies within the network. On the other hand, assuming self-similar traffic flows, any improvements in efficiencies through shaping are likely to be offset by the cost of relocating a large number of buffers to the periphery (vastly increasing overall buffer requirements, because there is less scope for sharing at the edge) and by performance penalties; instead, network architectures which exploit the full potential for multiplexing gains, are suggested. This is *not* to suggest that shaping and policing have no role in practical traffic management – indeed, these controls are essential to preserve fairness and stability by protecting contract abiding users from *atypically heavy* usage on the part of nonconforming connections.

4.4.5. Safe Operating Points

An important issue of concern in network operations and engineering is to determine thresholds for traffic levels such that (i) all QoS criteria are expected to be met, (ii) the network elements are stable in the presence of short-term load fluctuations and (iii) future growth are accounted for. When traffic levels consistently approach or exceed these levels, a servicing action needs to be invoked (such as adding capacity, or rearranging resources). As with other issues in traffic management, the setting of safe operating points depends (among other factors such as QoS objectives, bandwidth requirements, statistical multiplexing gains, link and trunk capacities, demand forecast, specific switch architecture, etc.) sensitively on traffic characteristics. In particular, using inappropriate models and methods can result in drastic under-engineering and poor service, or over-engineering and network inefficiencies as have already been experienced in actual broadband networks. See Figure 1 and the associated discussions in Section 2.1. More generally, taking into account the burstiness of traffic requires the notion of safe operating *regions* which are defined by combinations of traffic parameters (e.g., the FBM parameters) for which performance is acceptable.

4.4.6. Traffic Measurements

The primary bottleneck to applying in practice the many sophisticated traffic management methods that have been developed in theory over the years is our inability to specify the inputs required by such theory. Modeling the burstiness over many time scales of interest using extensions of traditional methods ("one time scale at a time") will typically require many parameters that need to be estimated on the basis of either special studies or regular operational measurements. Currently, switching systems and Operations Support Systems (OSSs) lack the capacity to collect, transport, process and store a large number of measurements; in fact, the only traffic measurement that is typically

SELF-SIMILAR TRAFFIC FLOWS 89

available is a rate or cell count measurement made over a coarse time scale on the order of 15-60 minutes. More frequent measurements, for example, based on more frequent SNMP queries may impose unacceptable measurement overheads.

The measurement and estimation problem is one of the motivations for seeking parsimonious, or compact, models of complex packet traffic phenomena. The FBM model can capture the burstiness over many time scales in aggregate data traffic with only three parameters. While this is a significant improvement, current switching systems still lack the capabilities to estimate the burstiness parameters (a and H). Credible settings of default ranges of the burstiness parameters can be estimated from the analysis of specialized, high-resolution traffic studies, on a per application or environment basis. In the longer term, it may be feasible and preferable to enhance the capabilities of switching systems to permit routine on-line estimation of burstiness parameters. In principle, the empirically observed self-similarity of our measured traffic can be exploited to reduce measurement overhead by noting that the value of H for a time series of counts over coarse time scales is the same as that obtained from high resolution traces (the value of a can also be inferred). Currently, some switching systems have the capability to report traffic counts over one second intervals during a data collection interval or an engineering period; and in principle, H can be estimated from such counts[21].

4.4.7. Connection Level Engineering

In high-speed networks that support *switched virtual connection* services (SVC), traffic management should seek to maintain network performance at both the cell and connection levels. The CAC function enables a logical separation of cell level and connection level engineering. In particular, engineering methods are required to ensure that sufficient resources are provisioned such that the blocking enforced by the CAC is at acceptable levels. Currently, several algorithms to calculate blocking performance in a multi-rate traffic environment such as ATM are available; these are valid in environments in which linear equivalent bandwidth descriptions are acceptable. The use of nonlinear admission algorithms, while important for realizing cell level efficiencies, complicates connection level engineering.

A second aspect of fractal traffic, namely call holding times that can span many time scales, also has considerable impact on connection level engineering. Traditional blocking analysis is predicated on assumptions that apply to voice networks, e.g., holding times of the order of 3 minutes. However, in a full service ATM environment, the holding times are expected to span many time scales, ranging from a few seconds for transaction-oriented applications, to many minutes and hours for video and data applications. Recent statistical analyses of measured network traffic have revealed *long holding times* (i.e., holding times that are consistent with infinite variance) not only in data networks [2,7,22] (e.g., ftp, telnet, www), but also in today's voice networks [4] (e.g., voice, fax, modem). Long holding times can severely impact connection level engineering[23]. The blocking levels predicted by traditional models apply in the "steady state", which is reached on time scales of the order of 5-10 average holding times. This means that when the holding time is of the order of an hour, actual performance will not converge to the theoretical values within an engineering period (15-180 minutes). Further, reattempts will appear correlated, resulting in higher than anticipated blocking. If the engineering periods are increased to account for the slow convergence to theoretical results, the underlying arrival process is no longer stationary, and time of day variations in the call arrival rate will have to be taken into account[24]. Note that the aforementioned problems associated with long holding time have already been observed in other data services and networks such as Internet access[25].

4.5. FUTURE WORK

Given the current state of the art in this field, virtually all the topics discussed in this chapter are still areas of active research. We identify a few of the issues that may be of more interest to the networking community:

- *Fast Generation of Self-Similar Traffic for Simulations:* We reviewed two techniques to generate self-similar traffic for simulation analysis of high-speed networks. We are aware of at least a dozen other promising techniques to generate self-similar traffic (e.g., the immigration-death process or $M/G/\infty$ model[26], wavelet-based methods[27]), though little is known so-far on the relative merits of these generation techniques, specifically on the speed vs. accuracy trade-offs of the various methods. This is a promising area of research, which can result in a standard library of generation techniques with clearly understood capabilities and limitations.

- *Fast Generation of Traffic with Arbitrary Variance-Time Characteristics:* More generally, there is a need for fast generation methods for traffic with more general variance-time characteristics than exactly self-similar traffic for e.g., superpositions of heterogeneous FBM traffic streams, variable-bit-rate (VBR) video which is asymptotically self-similar. A number of techniques are feasible in principle (e.g., generalized RMD, shaping of FBM traces, fractional ARIMA models) but once again, an understanding of the speed vs. accuracy trade-offs in each method is an area for further research.

- *Monitoring Networks:* Networks are currently monitored, for the most part, on the basis of coarse time scale measurements (15-60 minutes) which conveys information on the "quantity" of the traffic, but not its "quality" or burstiness. Additional research is needed to identify a small, sufficient set of statistics whose on-line estimation is feasible. As indicated in Section 4, the self-similarity of traffic can be exploited in principle to reduce the measurement overhead of too-frequent rate or count measurements.

- *Connection Level Management:* Methods that can be used to engineer in the presence of "fractal holding times" e.g., which can span a wide range of time scales, from milliseconds (e.g., ftp) through hours (remote login, videoconferencing) are required to address the issues identified in Section 4.8. Also, needed are methods for connection level engineering when nonlinear CAC algorithms are used to block connections.

- *Interactions Between Controls and Traffic:* The FBM model is valid when the action of flow controls on any one user is negligible. More generally, a variety of network controls can interact with traffic flows in complex ways, and the extent to which source characteristics (heavy-tailed *ON/OFF* periods, LRD of aggregate streams, etc.) are altered by controls is an important area for further research.

- *Fractal Descriptions of QoS Processes:* QoS processes, such as cell losses, are extremely bursty, and long-term average values (such as cell loss rates) are of limited value in describing such processes. Alternate characterizations (e.g., fractal dimensions) are feasible, but to this point, the identification of a feasible set of performance measures to describe highly bursty QoS processes has eluded a definitive quantification.

- *Chaotic Behavior in Networks:* In an application of chaotic models in a different vein, the investigation of the dynamical behavior of networks is an area of importance. Communications networks can be viewed as a complex ensemble of coupled non-linear dynamical systems which are capable of important and consequential dynamical systems behavior (e.g., bi-stability, oscillations, chaos) which are not considered in standard steady-state analysis. In particular, the interactions between controls at the network and application levels with self-similar traffic characteristics is of interest.

ACKNOWLEDGEMENTS

The authors would like to gratefully acknowledge the contributions of numerous colleagues in developing our current understanding of the modeling and management of self-similar traffic flows in networks, including: Arnie Neidhardt, K.R. Krishnan, Don Smith, Parag Pruthi, Gopal Meempat of Bellcore; Ilkka Norros of VTT Finland; Darryl Veitch of the Royal Institute of Technology, Stockholm; Wing-Cheong Lau of SBC TRI; and Onuttom Narayan of the University of California, Santa Cruz.

REFERENCES

1. W. E. LELAND, M. S. TAQQU, W. WILLINGER, and D. V. WILSON, "On the self-similar nature of Ethernet traffic (extended version)", *IEEE/ACM Transactions on Networking* **2**, pp. 1–15, 1994.

2. V. PAXSON and S. FLOYD, "Wide area traffic: The failure of Poisson modeling", *IEEE/ACM Transactions on Networking* **3**,

pp. 226–244, 1995.

3. J. BERAN, R. SHERMAN, M.S. TAQQU and W. WILLINGER, "Long-range dependence in variable-bit-rate video traffic", *IEEE Transactions on Communications* **43**, pp. 1566–1579, 1995.

4. D.E. DUFFY, A.A. MCINTOSH, M. ROSENSTEIN and W. WILLINGER, "Statistical analysis of CCSN/SS7 traffic data from working CCS subnetworks", *IEEE Journal on Selected Areas in Communications* **12**, pp. 544–551, 1994.

5. W. WILLINGER, S. DEVADHAR, A. HEYBEY, R. SHERMAN, M. SULLIVAN and J.R. VOLLARO, "Measuring ATM traffic cell-by-cell: Experiences and preliminary findings from BAGNet", preprint, 1996.

6. J. FEDER, *Fractals*, Plenum Press, New York, 1988.

7. W. WILLINGER, M. S. TAQQU, R. SHERMAN and D. V. WILSON, "Self-similarity through high-variability: Statistical analysis of Ethernet LAN traffic at the source level", *Proceedings of ACM SIGCOMM'95*, pp. 100–113, Cambridge, MA, 1995.

8. B.B. MANDELBROT, *The Fractal Geometry of Nature*, W.H. Freeman and Co., San Francisco, 1982.

9. I. NORROS, "A Storage Model with Self-Similar Input", *Queueing Systems* **16**, pp. 387–396, 1994.

10. A. ERRAMILLI, O. NARAYAN and W. WILLINGER, "Experimental queueing analysis with long-range dependent packet traffic", *IEEE/ACM Transactions on Networking* (to appear).

11. F. BRICHET, J.W. ROBERTS, A. SIMONIAN and D. VEITCH, "Heavy traffic analysis of a fluid queue fed by On/Off with long-range dependence", preprint, 1995.

12. N.G. DUFFIELD and N. O'CONNELL, "Large deviation and overflow probabilities for the general single-server queue, with applications", *Proceedings of the Cambridge Philosophical Society* (to appear).

13. W.-C. LAU, A. ERRAMILLI, J.L. WANG and W. WILLINGER, "Self-similar traffic generation: The random midpoint displacement algorithm and its properties", *Proceedings of the ICC '95*, pp. 466-472, Seattle, WA, 1995.

14. B.B. MANDELBROT, "Long-run linearity, locally Gaussian processes, H-spectra and infinite variances", *International Economic Review* **10**, pp. 82–113, 1969.

15. P. PRUTHI, *An application of chaotic maps to packet traffic modeling*, PhD thesis, Royal Institute of Technology, Stockholm, Sweden, 1995.

16. M. CSENGER, "Early ATM Users Lose Data", *Communications Week*, May 16, 1994.

17. C.L. HWANG and S.-Q. LI, "On input state space reduction and buffer noneffective region", *Proceedings of IEEE Infocom '94*, pp. 1018–1028, 1994.

18. A. ELWALID and D. MITRA, "Effective bandwidth of general Markovian traffic sources and admission control of high-speed networks", *IEEE/ACM Transactions on Networking* **1**, pp. 329–343, 1993.

19. R. GUERIN, H. AHMADI and M. NAGHSHINEH, "Equivalent capacity and its application to bandwidth allocation in high-speed network", *IEEE Journal Selected Areas in Communications* **9**, pp. 968–981, 1991.

20. A. ERRAMILLI and A.L. NEIDHARDT, "Roles of Shaping and Policing in Traffic Management", preprint, 1996.

21. A. ERRAMILLI and J.L. WANG, "Monitoring packet traffic levels", *Proceedings of the IEEE Globecom '94*, pp. 274–280, San Francisco, CA, 1994.

22. M.E. CROVELLA and A. BESTAVROS, "Explaining world wide web traffic self-similarity", preprint, 1995.

23. A. ERRAMILLI, E.H. LIPPER and J.L. WANG, "Some performance considerations for mass market broadband services", *Proceedings of the 1994 1st International Workshop on Community Networking*, pp. 109–116, Millbrae, CA, 1994.

24. A. ERRAMILLI, J. GORDON and W. WILLINGER, "Applications of fractals in engineering for realistic traffic processes", *The Fundamental Role of Teletraffic in The Evolution of Telecommunications Networks (Proceedings of ITC-14, Antibes Juan-les-Pins, France, June 1994)*, J. Labetoulle and J.W. Roberts (Eds.), pp. 35–44, Elsevier, Amsterdam, 1994.

25. "Bell Companies Assail AT&T's Internet Plan", *New York Times*, February 29, 1996.

26. D.R. COX, "Long-range dependence: A review", *Statistics: An Appraisal*, H.A. David and H.T. David (Eds.), pp. 55–74, Iowa State University Press, 1984.

27. P. FLANDRIN, "Wavelet analysis and synthesis of fractional Brownian motion", *IEEE Transactions on Information Theory* **38**, pp. 910–917, 1992.

CHAPTER 5
A NEW TRAFFIC CONTROL MECHANISM FOR CONTINUOUS MEDIA COMMUNICATIONS

Frank Ball, David Hutchison and Demetres Kouvatsos

5.1. INTRODUCTION

Distributed multimedia applications will require support for continuous media traffic, ie digitised audio and video. Continuous media place a high demand on both the network and the workstation, requiring not only high throughput, but also timely delivery. Therefore communications performance and predictability are important issues, particularly if these applications are to be provided with a guaranteed Quality of Service (QoS).

In order to provide a guaranteed end-to-end QoS to an individual connection, resource allocation will be required at every node in the network, along the path of that connection. Resource reservation protocols have been proposed for use in the internet environment,[1] and end-to-end methods of bounding delay in packet switched networks have been developed.[2,3] However, these cases are based on the assumption that either the network comprises homogeneous nodes, or that packets remain unchanged as they migrate through the network. For a network comprising heterogeneous nodes, the differing resource allocation mechanisms of individual subnetworks, and the effects of fragmentation/packetisation will need to be considered.

An Enhanced Network Layer Architecture (ENLA) has been proposed,[4,5] to support guaranteed services to continuous media across multi-hop heterogeneous networks. This architecture is intended to supplement a more general communication architecture, viz. the Quality of Service Architecture (QoS-A), which provides a framework for QoS specification and resource control, over all architectural layers, from the application platform to the network.[6] In addition to mechanisms such as resources reservation, resource allocation and call admission control, ENLA introduces two new mechanisms which aid the resources reservation protocol by matching the traffic characteristics and QoS requirements of the source to the performance characteristics of individual sub-nets, viz temporal mapping and QoS mapping.

Although logically separate functions, temporal mapping and QoS mapping will need to work co-operatively, therefore for practical purposes they may be combined into a single process. The role of temporal and QoS mapping is to match the temporal characteristics of the source with the temporal characteristics of the network. The main objective is to maximise the efficiency of bandwidth allocation whilst at the same time ensuring that the QoS requirements of the connection are respected. In order to achieve this it may be necessary to fragment packets and space out the transmission of these fragments over a period of time. In this case it is essential that the delay requirements of the connection are not exceeded.

In this paper a queueing model has been developed concerning the temporal and QoS mapping fragmentation process, which may be applied to both ATM-like and FDDI-like networks. This model abstracts out some of the detail involved in the fragmentation process into a single queue. The delay and loss probability distributions for the individual mappings can then be obtained by solving this simple queueing model. Of particular interest is the determination of the percentile delay which is useful for bounding both jitter and end-to-end delays. A fundamental aspect in the solution process of the queueing model is the characterisation of its interarrival and service (transmission) time distributions.

Measurements of actual traffic or service times are generally limited and so only few parameters can be computed reliably. Typically, only the mean and variance can be relied upon. In this case, the choice of distribution which implies least bias (ie., introduction of arbitrary and, therefore, false assumptions) is that of the Generalised Exponential (GE) distribution.[7,8] In this paper, the proposed queueing model is approximated by a stable infinite capacity GE/GE/1 queue with a single server and GE-type interarrival and service time distributions. The exact response time distribution and percentile delays of this queue are determined using probabilistic arguments.

The operations of the temporal mapping and QoS mapping functions are introduced in Section 5.2, where the temporal structure of continuous media traffic and the temporal characteristics of a number of different subnetworks are identified. In Section 5.3, the effects of packet fragmentation, for the purpose of temporal mapping, is discussed and a queueing model of the fragmentation process is presented. Moreover, applications of this queueing model to both ATM and FDDI networks are discussed and the main performance parameters are identified. Section 5.4 introduces the GE distribution and carries out theoretical work to determine closed-form expressions for the response time distribution and percentile delays in a stable GE/GE/1 queue. A specific application example of temporal mapping on to an FDDI network is demonstrated in Section 5.5. Conclusions and comments on further work follow in Section 5.6.

5.2. TEMPORAL AND QOS MAPPING

The role of QoS mapping is to match the QoS requirements of the connection to the services which are offered by the underlying subnetwork. The QoS mapping function will choose the most appropriate service offered by the subnetwork in order to meet the QoS requirements of the connection.

Traffic flows through a network have been shown to comprise a multi-layer temporal structure.[9] This structure may change as it passes through the network, and a transformation in structure may occur as traffic is passed between protocol layers, or between different subnets in a heterogeneous network. The capacity offered by a particular subnetwork may also possess temporal properties, ie the subnetwork may give a variable response time to the transmission of data units. The traffic flow and the subnetwork may also operate at a different temporal granularity, and therefore mapping between the two may be required. In this case fragmentation of data units may be necessary in order to match the temporal granularity of the incoming traffic to that of the subnetwork. This mapping may alter the original traffic structure, therefore details of the resulting traffic structure must be reported to the next node along the path of the connection.

Temporal mapping will match the temporal structure of the traffic onto the temporal characteristics of the subnetwork. Generally this mapping will need to be carried out before any call admission test can be applied. The temporal mapping will also report any resulting change in temporal structure to the next stage of the resource reservation process.

5.2.1. Traffic Characterisation

Filipiak [9] identifies three layers in the temporal structure of traffic flows: the call layer, the burst layer and the packet layer. Each layer is characterised by its time scale, which is defined by the mean interarrival time of the layer entities. The flow of entities can be described by two random variables, the interarrival time of the entity (I) and the entity size (L). Therefore traffic can be fully categorised by the following set of parameters:-

- I^C : Time between calls.
- L^C : call duration.
- I^B : Inter-burst time.
- L^B : Burst length or duration.
- I^P : Interpacket time.
- L^P : Packet length.

For the purpose of QoS and temporal mapping, only the burst and packet level parameters are required. However the call level parameters may be useful to other functions, eg. advanced resource reservation.

Generally, each of the parameters in this set will represent a random variable. A random variable may be described by a pdf/pmf together with certain parameters of this pdf/pmf, eg mean, variance. However, in most cases it is unlikely that a network user will be able to provide full statistical information about its traffic source. Therefore the QoS and temporal mapping process must be able to function with whatever information is made available.

Generally, where only partial information is provided, worst case assumptions may need to be made. As an example, consider the case where only the maximum, mean, and minimum values of a random variable are given. The pmf, say A(t), of a distribution for that random variable, which for those three values, will give a worst case variance, is given by the following [10] :-

$$A(t) = \begin{cases} \dfrac{\text{mean-max}}{\text{max-min}} & t = \max \\ 1 - \dfrac{\text{mean-max}}{\text{max-min}} & t = \min \\ 0 & \text{otherwise} \end{cases}$$

In this paper, unless otherwise stated, we assume that the user can provide maximum, minimum and average values for both the packet interarrival time and packet length.

5.2.2. Network Temporal Characteristics

Certain types of network provide a transmission channel which exhibits temporal characteristics. Packets awaiting transmission will experience an access delay before being transmitted across the network. This access delay, which is the response time of the network, is mainly influenced by the time between successive transmission opportunities (T^o) at the access point to the network.

At each transmission opportunity a station may transmit a given amount of data (A^x), the maximum size of which is determined by the rules of the particular network access protocol. The temporal characteristics of a number of network types are given below.

A NEW TRAFFIC CONTROL MECHANISM

Network	T^0	A^x
FDDI-II (Isoch) DQDB PA	Fixed (125µs)	Fixed (1,2, ..,N octets)
FDDI (synchronous)	Bounded (Min, Ave, Max)	Pre-allocate max. Multiple-frames.
Token Ring	Bounded (Min, Max)	One frame of up to maximum length
ATM	Arbitrary, Pre-defined	Maximum number of cells
CSMA/CD	Non-Deterministic	One frame of up to maximum length

Table 5.1. Network Temporal Characteristics

In some cases, eg FDDI, A^x may be determined by negotiation, whilst in others, eg a Token Ring, A^x is influenced by the size of the frame. The transmission time (t) of A^x will generally be much less than T^0 and is determined by both A^x and the transmission rate of the network.

5.2.3. Mapping Traffic Structure to Network Characteristics

This mapping will take place at the access point to the subnetwork. Therefore, when choosing the appropriate mapping strategy, the main QoS parameters of concern will be the delay and loss which could be experienced by packets awaiting access to the network. There are two general cases of subnetworks which need to be considered :-
    1) FDDI-like subnetworks which allow a variable length packet, and where the transmission interopportunity time is determined by the subnetwork.
    2) ATM-like subnetworks which use a fixed length packet.
Each of these will require a different temporal mapping strategy.

5.2.3.1. Mapping to FDDI-like subnetworks.
Consider packets from a number of sources that arrive at the access point to a subnetwork and which join a queue before being transmitted. The combined average packet interarrival time must be greater than T^0_{ave} for the network, otherwise the size of the transmission queue would continue to grow without limit resulting in eventual buffer overflow for a finite buffer.

However, in most cases the average interarrival time of an individual connection (I^p_{ave}) will be much greater than T^o_{ave}, and in this case two options are possible:-
  1) Transmit a full packet at the first transmission opportunity, then ignore all subsequent transmission opportunities until the next packet becomes available.
  2) Fragment the packet into a number of smaller units and transmit these units over a number of transmission opportunities.

If the first option is chosen it may result in an inefficient utilisation of bandwidth. However, if the second option is chosen, care must be taken to ensure that the QoS requirements of the connection are met.

Choosing to fragment incoming packets leads to an increase packet transmission time, and hence an increase in the access delay onto the network. The greater the number of fragments (n), the greater the access delay, but also the greater the efficiency of bandwidth allocation. Therefore, the goal should be to maximise n, within the delay constraints. Determining the delay for a given n will involve the solution of some form of queueing model.

5.2.3.2. Mapping to ATM-like subnetworks.
This type of network uses a small fixed length packet or cell, therefore in most cases incoming packets will need to be fragmented. The number of fragments (n) will be determined by the length of the incoming packet (L), and the payload size of the cell (f); this is given by :-

$$n = \left\lceil \frac{L}{f} \right\rceil$$

In this case it is necessary to choose the cell transmission rate (a) for the connection. At this point there are two options :-
  1) Convert the characteristics of the incoming traffic directly into the required ATM cell level parameters, ie peak cell rate, mean cell rate and mean burst length. Call admission will be then based on these parameters, and a will be equal to the peak cell rate. Since all cells will be transmitted at a peak rate which is equivalent to that of the original source, delay at the packet queue will be minimal.
  2) Choose a to be equivalent to a rate between the peak and mean rates of the source and thereby produce a smoother cell level traffic profile. This will of course impose a delay at the packet queue

In the first case the peak cell rate (R_p) is given by: $\dfrac{1}{R_p} = \left\lfloor \dfrac{\Delta I^p_{min}}{n_{max}} \right\rfloor \dfrac{1}{\Delta}$

A NEW TRAFFIC CONTROL MECHANISM 103

where D = cell rate of the link and $n_{max} = \left\lceil \dfrac{L^p_{max}}{f} \right\rceil$.

The mean cell rate (R_a) may be obtain in a similar manner using I_p .ave and n_{ave}. Burst length dimensions however, may be more difficult to estimate.

In the second case the lower the value of a (where $R_a < a <= R_p$)the smoother the cell level traffic profile and hence the greater the efficient in bandwidth utilisation. However, delays will increase as a reduces, therefore a should be minimised within the delay constraints. As with the FDDI case this will involve the solution of some form of queueing model. The next section describes a queueing model which may be used for both the FDDI and ATM cases.

5.3. TEMPORAL MAPPING QUEUEING MODEL

The process of fragmentation may be carried out in a number of ways, depending upon particular implementations. Packets may be fragmented within host memory, with fragments being transferred individually across the I/O bus into a buffer on the network adaptor. Alternatively packets may be transferred across the I/O bus in their entirety and fragmented onboard the network adaptor.

In order to minimise the complexity of the temporal mapping process, a relatively simple delay estimation model is required. Since it is delay at the packet level which is of greatest interest, a model that estimate delay directly at this level will be the most suitable.

We have developed a queueing model for the temporal mapping fragmentation process, which abstracts out much of the detail of the underlying implementation, and allows packet delay to be estimated by solving a single queue.

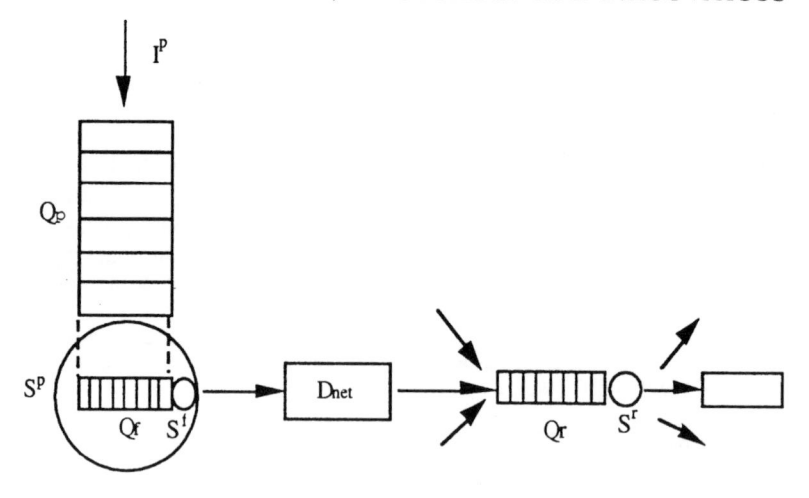

Fig. 5.1. Fragmentation Queueing Model

Figure 5.1 shows a general model for the fragmentation, transmission and reassembly of packets across a network. Specific applications of this model may omit certain sections, eg. re-assembly, as required. Also in certain cases it may be necessary to add implementation specific processing delays.

The incoming packets join the packet queue (Q_p) with an inter-arrival time of I^p. When a packet at the head of Q_p goes into service it is instantaneously formed into n equal size fragments which are immediately loaded into the transmission queue (Q_f). Q_f is then served until empty with a service time of S^f. Between the time of the initial loading of Q_f and the time that Q_f becomes empty, no further fragments are allowed to join the Q_f. This model can be considered as a two dimensional queuing model in which the time taken to empty Q_f becomes the service time of Q_p (S^p).

For networks such as FDDI, S^f represents T^0 which will generally be a random variable. Alternatively for an ATM network S^f represents the inter-cell transmission time of a particular VC and will generally be a multiple of the minimum inter-cell transmission time of the ATM link and therefore constant. Where S^f is a random variable, subscripts are used to denote a particular instance of S^f eg. $S^f_{j,i}$ = the service time as experienced by the ith fragment of packet j

A NEW TRAFFIC CONTROL MECHANISM

The timing relationships of the system are given below and illustrated in figure 5.2.

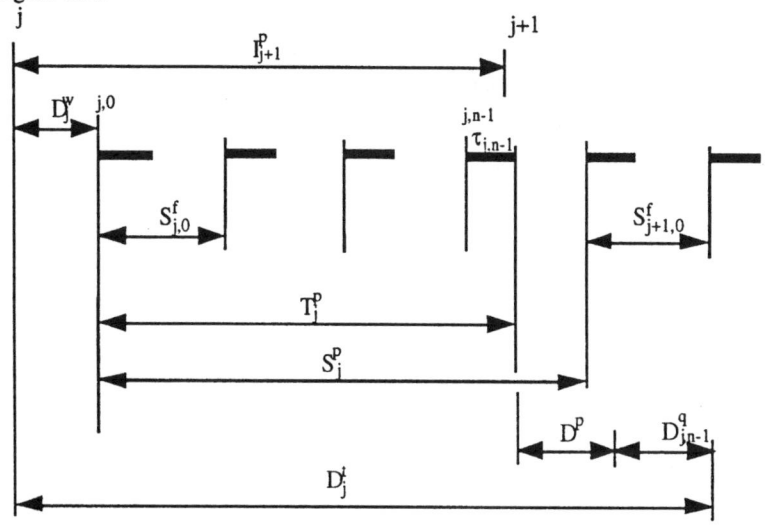

Fig. 5.2. Fragmentation Timing Diagram

j = the arrival time of the jth packet and I^p_{j+1} = the inter-arrival time between the jth and the j+1th packet.

D^w_j = The waiting time experienced by packet j. Even when the jth packet arrives to find an empty queue with no packets in service, there may still be a waiting time due to the slotted nature of the system. In this case D^w_j will be bounded by S^f_{max}.

S^p_j = the service time of the jth packet and is given by :-

$$\sum_{i=0}^{n} S^f_{j,i} \quad \text{alternatively, if } S^f \text{ is a constant } \forall\ j,i,\ S^f = nS^f$$

where $S^f(t)$ is the pdf/pmf of S^f, a pdf/pmf for S^p ($S^p(t)$) can be formed by the nth fold convolution of $S^f(t)$

T^p_j = the time taken to transmit the jth packet and is given by :-

$$\sum_{i=0}^{n-1} S_{j,i}^f + \tau_{j,n-1}$$

alternately, if S^f is a constant $\forall$ j,i, $T^p = (n-1)S^f + \tau_{j,n-1}$

D_j^t = the transit delay experiences by the jth packet and is given by :-

$$D_j^w + T_j^p + D^p + D_{j,n-1}^q$$

Where D^p = A fixed delay due to the latency in the network
and $D_{j,n-1}^q$ = A variable delay due to queueing delays within the network.

For shared medium networks, eg FDDI, there will be no internal network queuing delay; however, a delay may still be experienced at the receiving queue.

Assuming Q_p to be of infinite size then for stability the utilisation (ρ) must be less than unity. This requirement is given by the following :-

$$\rho = \frac{S_{ave}^p}{I_{ave}^p} < 1,$$

where $S_{ave}^p = nS_{ave}^f$, $n \in N$.

Therefore given that I_{ave}^p will be supplied known from the characteristics of the incoming traffic, the temporal and QoS mapping process will need to chose a combination of n and S_{ave}^f which complies with the above constraint.

In order to maximise the efficiency of bandwidth allocation, the QoS and temporal mapping process should choose a combination of n and S_{ave}^f which maximises ρ within the required delay constraints. Therefore, once an initial combination of n and S_{ave}^f has been chosen, further tests will be required to ensure that the QoS requirements will be met. The sequence of operations of the QoS and temporal mapping process will therefore be as follows:-

1) Given I_{ave}^p choose a combination of n and S_{ave}^f which maximise

ρ subject to the constraint $\rho < 1$

A NEW TRAFFIC CONTROL MECHANISM

2) Estimate the delay and loss at Q_p for this combination.
3) if the QoS requirements of the call cannot be met, make adjustments to the combination and repeat step 2.

Delay/loss will be estimated by using the queuing model described above together with information provided of the arrival characteristics of the traffic, and knowledge of the temporal properties of the individual sub-network. The resulting queueing model will approximate a single queue of a particular type. eg M/G/1, D/G/1, G/G/1 etc.

Although both delay and loss are of importance, in this paper we will consider only delay, and assume buffer sizes are sufficiently large to prevent loss in all cases. The mean delay through the system may be of interest, although the most important parameter is the q-percentile response time. This will a provide an upperbound on delay with an associated probability, and will therefore be useful for bounding jitter and end-to-end delay. In the next section the method for estimating the q-percentile response time is presented.

5.4. THE GE-TYPE PERCENTILE RESPONSE TIME

This section introduces the GE distribution and performs exact analysis in order to determine the response time distribution and percentile delay of a stable GE/GE/1 queue.

5.4.1. The GE Distribution

The GE distribution is of the form

$$F(t) = P(X \le t) = 1 - \tau e^{-\sigma t}, t \ge 0,$$

where

$$\tau = 2/(C^2 + 1),$$

$$\sigma = \tau v,$$

X is a mixed-time random variable (rv) of the interevent-time, while $1/v$ is the mean and C^2 is the squared coefficient of variation (SCV) of rv X (c.f., Fig 5.3).

$$1-\tau = \frac{C^2-1}{C^2+1}$$

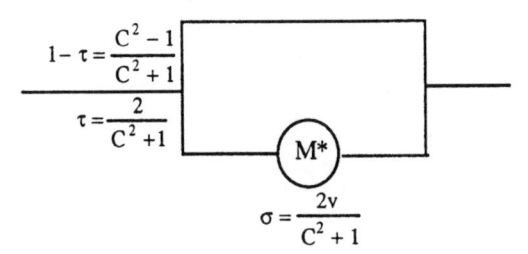

$$\tau = \frac{2}{C^2+1}$$

M*

$$\sigma = \frac{2v}{C^2+1}$$

Fig. 5.3. The GE distribution with parameters τ and σ.

For $C^2 > 1$, the GE model is a mixed-time probability distribution and it can be interpreted as either

    (i) an extremal case of the family of two-phase exponential distributions (e.g., Hyperexponential-2 (H_2)) having the same v and C^2, where one of the two phases has zero service time; or

    (ii) a bulk type distribution with an underlying counting process equivalent to a Compound Poisson Process (CPP) with parameter $2v/(C^2+1)$ and geometrically distributed bulk sizes with mean $=(C^2+1)/2$ and SCV $=(C^2-1)/(C^2+1)$ given by

$$P(N_{cp}=n) = \begin{cases} \sum_{i=1}^{n}\frac{\sigma^i}{i!}e^{-\sigma}\binom{n-1}{i-1}\tau^i(1-\tau)^{n-i}, & \text{if } n \geq 1, \\ e^{-\sigma}, & \text{if } n=0, \end{cases}$$

where N_{cp} is a Compound Poisson (CP) rv of the number of events per unit time corresponding to a stationary GE-type interevent rv.

The GE distribution is versatile, possessing pseudo-memoryless properties which make the solution of many GE-type queueing systems and networks analytically tractable (e.g.[8,11,12,13]). Moreover, it has been experimentally established that the GE model, due to its extremal nature, defines performance bounds over corresponding solutions based on two-phase distributions with the same two moments as the GE. The GE distribution is completely characterised in terms of its first two moments and it can be interpreted as a maximum entropy solution (c.f.,[14]), subject to the constraints of normalisation, discrete-time zero probability and expected value. In this sense, it can be viewed as the least biased distribution

---

* M denotes an exponential distributions

A NEW TRAFFIC CONTROL MECHANISM 109

estimate, given the available information in terms of the constraints. For $C^2 <1$, the GE distributional model (with $F(0)<1$) cannot be physically interpreted as a stochastic model. However, it can be meaningfully considered as a pseudo-distribution function of a flow model approximation of an underlying stochastic model (with $C^2 <1$) in which negative branching pseudo-probabilities (or weights) are permitted. In this sense, all analytical GE-type exact and approximate results obtained for queueing systems and networks when $C^2 <1$, can also be used - by analogy - as useful heuristic approximations when $C^2 <1$ as long as they satisfy basic queueing theoretic constraints. Note that the utility of other improper two-phase type distributions with $C^2 <1$ has been proposed in the field of systems modelling by various authors (e.g.[15,16]).

5.4.2. The GE-Type Response Time Distribution

Consider a stable GE/GE/1 queue with infinite capacity, a single server and GE-type interarrival and service time distributions.
Let

    λ be the mean arrival rate

    μ be the mean service rate

    C_a^2 be the interarrival time SCV, and

    C_s^2 be the service time SCV.

By applying probabilistic arguments and using the bulk interpretation of the GE-type distribution the following theorem holds:
Theorem 1: The response time distribution of an individual job in a stable $GE(\lambda,C_a^2)/GE(\mu,C_s^2)/1$ queue is of GE-type, namely

$$R(t)=P(W\leq t)=1- \tau_w e^{-\zeta \tau_w t}, \quad t \geq 0,$$

with parameters

$$\zeta = \frac{\tau_a \tau_s \mu(1-\rho)}{\tau_s(1-\rho\tau_a)+\rho\tau_a},$$

$$C_w^2 = \frac{2-\tau_w}{\tau_w},$$

where W is a mixed-time rv of the response time,

$$\tau_w = \frac{\tau_s(1-\rho\tau_a)+\rho\tau_a}{\tau_s(1-\tau_a)+\tau_a},$$

$$\tau_a = \frac{2}{1+C_a^2},$$

$$\tau_s = \frac{2}{1+C_s^2},$$

and $\rho=\lambda/\mu$.

#

The proof can be seen in Appendix I.

The q-percentile value of the response time, $\pi_w(q)$, can be computed by making use of Theorem 1 and is given via the following corollary.

<u>Corollary 1</u>: The q-percentile of the response time, $\pi_w(q)$, of a stable $GE(\lambda,C_a^2)/GE(\mu,C_s^2)/1$ queue is given by

$$\pi_w(q) = -\frac{1}{\zeta\tau_w}\ln\left(\frac{1-q}{\tau_w}\right)$$

#

The proof can be seen in Appendix II.

5.5. AN EXAMPLE OF TEMPORAL MAPPING ONTO FDDI

This example considers the temporal mapping of a video source onto an FDDI network. Packets from the video source are presented to the network via a transport layer protocol which has been optimised for continuous media traffic[17]. This protocol employs rate based flow control in which the rate at which packets are submitted to the network is controlled by a timing mechanism. Generally, the timer operates at a rate equal to the periodicity of the source, eg. 40ms for 25 frames/s video, which in this example also equals I_{ave}^p.

The timed token protocol of FDDI guarantees the average rotation time to be bounded by the Target Token Rotation Time (TTRT), and the maximum rotation time to be bounded by 2*TRTT. The minimum possible rotation time is determined by the latency of the ring. Therefore for an

A NEW TRAFFIC CONTROL MECHANISM

FDDI network, in a worst case:-

$$S^f_{max} = 2*TTRT$$

$$S^f_{ave} = TTRT$$

$$S^f_{min} = \text{Ring Latency}$$

The actual distribution function of token rotation times $S^f(t)$ is not known, but for a worst case distribution this can be formed by the method given in section 5.2.1. Assuming only max, ave and min values of the traffic characteristics have been supplied, the same method can be used to form a distribution function for the packet arrival process. An initial value for n can then be obtained by the following :-

$$\text{If } I^p_{ave} \text{ MOD } S^p_{ave} = 0 \quad n = \frac{I^p_{ave}}{S^f_{ave}} -1 \quad \text{else} \quad n = \left\lfloor \frac{I^p_{ave}}{S^f_{ave}} \right\rfloor$$

Generally the video packets will be of variable length packets. However, since n is chosen to optimally match the temporal granularity of I^p with that of T^o, packet length will have no effects on the choice. Each packet will be divided into n approximately equal sized fragments, the size of which may vary from packet to packet. Bandwidth allocation for this individual connection (A^x_k) will be based on the maximum possible size of fragment given by :-

$$f_{max} = \left\lceil \frac{L^p_{max}}{n} \right\rceil$$

Using the first two moments of both the arrival time distribution, and the service time distribution which has been obtained by the n fold convolution of $S^f(t)$, the required q-percentile delay will be calculated by use of the formula given in section 5.4. If this delay not acceptable n will be decremented and the above process will be repeated.

To validate use of the fragmentation queueing model, we have considered two cases of temporal mapping onto an FDDI network:-

1) Packet arrival process poisson with $l=I^p_{ave}$ =40ms, TTRT=10ms, Latency=3ms.

2) Packet arrival process poisson with $l=I^p_{ave}$ =40ms, TTRT=5ms, Latency=1ms.

The 99-percentile delays where calculated for n = 1, 2, 3 in case 1, and n = 1, 2, 3, 4, 5, 6, 7 in cases 2.

The 99-percentile delays for both cases, over the same range of values of n, were also obtained by simulation. The simulation models the fragmentation of packets within host memory, and the transfer of the fragments across the i/o bus into the transmission queue of the network adaptor. Unlike the queueing model presented in section 5.3, the simulation model closely matches the fragmentation process as it could be carried out in a real system. The only major assumption is that the host can support real time scheduling, and that the fragmentation of individual packets will not be interrupted. The rotation times were generated according to a worst case distribution based on S^f_{max}, S^f_{ave} and S^f_{min}.

The values for 99-percentile delay, obtain by both calculation and simulation, are given below:-

I^p_{ave} = 40ms		C^2_a = 1		99-percentile Delay (ms)	
n	S^f_{ave} (ms)	r	C^2_s	Calculation	Simulation 99% CI
1	10	0.250	0.700	53.600	39.6 - 41.2
2	20	0.500	0.350	130.164	114.6 - 118.9
3	30	0.750	0.175	336.065	310.6 - 342.8

Table 5.2. 99-percentile Delays for Case 1

I^p_{ave} = 40ms		C^2_a = 1		99-Percentile Delay (ms)	
n	S^f_{ave} (ms)	ρ	C^2_s	Calculation	Simulation 99% CI
1	5	0.125	0.8000	24.160	16.5
2	10	0.250	0.400	45.583	36.1 - 37.3
3	15	0.375	0.200	71.33	62.9 - 65.1
4	20	0.500	0.1000	108.859	101.1 - 106.1
5	25	0.625	0.0500	171.406	164.3 - 180.3
6	30	0.750	0.0250	296.338	292.1 - 321.3
7	35	0.875	0.0125	699.089	637.0 - 915.0

Table 5.3. 99-percentile Delays for Case 2

It can be seen that in case 1 the predicted delays are greater than the upper confidence limit of the simulation results for all values of n. This is also true in case 2 for n < 5. For the remaining three values of n, the

A NEW TRAFFIC CONTROL MECHANISM 113

predicted delay lies within the 99% confidence region of the simulation results. However, re-evaluating the figures using a 20% level of confidence (see table 5.4), shows that at this level significance the predictions underestimate delay for n = 5,6 and 7.

n	99-percentile Delay (ms) Calculation	Simulation 60% CI
5	171.406	171.65 - 172.90
6	296.338	305.52 - 307.87
7	699.089	764.63 - 786.97

Table 5.4. 99-percentile Delays at 20% Confidence Level

For a TTRT of 10ms, reducing the synchronous bandwidth allocation by one third, ie choosing n = 3, results in approximately a six fold increase in the value of the 99-percentile delay. On the other hand, for a TTRT of 5ms, reducing the synchronous bandwidth allocation by a similar amount results in only approximately a three fold increase. This difference is due to the fact that the main influence on delay is the utilisation r and not the number of fragments n.

Within the simulation model a fixed delay was use to represent the transfer time of individual fragments across the system bus between the host memory and the network adaptor. It was found that provided this delay is less than one half of the minimum token rotation time, then it has no effect on the outcome of the simulation.

5.6. CONCLUSION AND FUTURE WORK

We have presented a performance study of a new traffic control mechanism, viz. the temporal mapping function, whose role is to match the temporal structure of continuous media traffic to the temporal characteristics of the network. A description of the temporal mapping processes has been given, followed by a detailed analysis of the temporal mapping queueing model.

An exact analysis which determines the response time distribution and the q-percentile delay of a stable GE/GE/1 queue has been presented. An example of the temporal mapping of a video source onto an FDDI network has been given, in which the 99-percentile delays of the temporal mapping queue are approximated by employing a stable GE/GE/1 queue. Finally the results of a simulation, designed to verify the temporal mapping queueing model, are presented.

A comparison between the values for 99-percentile delay derived by calculation with those obtained by simulations has shown that for the two cases which were considered, the model overestimates delay for $n \leq 4$, and underestimates delay for all higher values.

It has been shown that for an FDDI network, temporal mapping can provide improvements in the efficiency of bandwidth allocation, provided the consequential increase in delay is acceptable. Even where such improvements are not required it will still be necessary to estimate the delay for n=1, therefore the temporal mapping queueing model may still prove useful in this case.

Future work will consider cases where the packet arrival process is not Poisson; these will include deterministic arrivals, and cases of $C_a^2 \gg 1$ which could most likely occur where the FDDI station is a gateway receiving packets from other networks. The effectiveness of this model in the ATM case will also be investigated.

ACKNOWLEDGEMENTS

The work presented in this paper has been partly carried out within the Quality of Service Architecture (QoS-A) project, which is funded as part of the UK EPSRC Specially Promoted Programme in Integrated Multiservice Communication Networks (GR/H77194) in co-operation with GDC (formerly Netcomm Ltd). We also gratefully acknowledge funding from BT Labs as part of their University Research Initiative to investigate QoS management of multiservice networks. In addition, the work is partly supported by UK EPSRC grants GR/H18609 and GR/K/67809 on the performance modelling of computer communication networks.

References

1. D. J. MITZEL, D. ESTRIN, S. SHENKER and L. ZHANG. "An Archectural Comparison of ST-11 and RSVP", INFOCOM 94, Vol 2, 716-725 (June 1994) Toronto, Canada.
2. D. FERRARI and D. C. VERMA. "A Scheme for Real-Time Channel Establishment in Wide-Area Networks.", IEEE JSAC, 8, 3, 368-379 (April 1990).
3. J. KUROSE. "On Computing Per-session Performance Bounds in High Speed Multi-hop Computer Networks.", Performance Evaluation Review, 20, 1, 128-139 (June 1992) .
4. F. BALL and D. HUTCHISON, "An Architecture For SupportingGuaranteed Services in Heterogeneous Networks", EFOC & N 95, 124-127 (27-30 June 1995) Brighton, UK.
5. F. BALL and D. HUTCHISON, "Supporting Quality of service guarantees across Heterogeneous Lans", in Proceedings of UKPEW 95, M. Merabti, M.Carew and F. ball (Eds), (Springer-Verlag, 1995), 94 - 108.
6. A. CAMPBELL, G. COULSON and D. HUTCHISON. "A Quality of Service Architecture", Computer Communication Review, 24, 2, 6-27 (April 1994).

7. D.D. KOUVATSOS, "A Maximum Entropy Analysis of and the G/G/1 Queue at Equilibrium", J. Opl. Res. Soc., 39, 183-200 (1988).
8. D.D. KOUVATSOS, "Entropy Maximisation and Queueing Network Models", Annals of Operations Research, Special Issue on Queueing Networks, 48, 63-126 (1994).
9. J. FILIPIAK, "Structure of Traffic Flow in Multiservice Networks", INFOCOM 88, 425-429 (March 1988) New Orleans, USA.
10. H. SAITO and K. SHIORMOTO. "Dynamic Call Admission Control in ATM networks", IEEE JSAC, 9, 7, 982-989.
11. D.D. KOUVATSOS, "A Universal Maximum Entropy Algorithm for the Analysis of General Closed Networks", in Computing Networking and Performance Evaluation, IFIP WG 7.3, T. Hasegawa, et al (Eds.), (North Holland, 1986), 113-124.
12. D.D. KOUVATSOS and N. TABET-AOUEL, "Product-Form Approximations for an Extended Class of General Closed Queueing Networks", Performance '90, IFIP WG 7.3 and BCS, P. King et al (Eds.), (North-Holland,1990), 301-315, .
13. D.D. KOUVATSOS and N.TABET-AOUEL, "An ME-Based Approximation for Multi-Server Queues with Preemptive Priority", European Journal of Oper. Res., 77, 496-515 (1994).
14. E.T. JAYNES, "Prior probabilities", IEEE Trans. Syst. Sci. Cybern, SSC-4, 227-241 (1968).
15. C. SAUER, "Configuration of Computing Systems: An Approach Using Queueing Network Models", PhD Thesis, University of Texas, 1975
16. S. NOJO and H. WATANABE, "A New Stage Method Getting Arbitrary Coefficient of Variation by Two Stages", Trans. IEICE, 70, 33-36 (1987).
17. F.J.GARCIA. "Continuous Media Transport and Orchestration Services" PhD Thesis (May 93) Lancaster University.
18. D.D. KOUVATSOS, P. GEORGATSOS, and N. XENIOS, "On the Analysis of the GE/GE/1 Queue", Research Report DDK/17, Department of Computing, University of Bradford, January 1989.
19. L. KLEINROCK, Queueing Systems, Vol.1: Theory, (John Wiley and Sons, Inc., 1975), ISBN 0-471-49110-1.

Appendix I: Outline of proof for Theorem 1.

The total time taken for a whole bulk to be served in a stable $GE(\lambda, C_a^2)/GE(\mu, C_s^2)/1$ queue is equal to the random sum of GE service times over a geometric counter with parameter

$\tau_a = 2/(1+C_a^2)$. Thus, the total service time of a bulk is of GE-type, namely $GE(\tau_a \mu, C_{sb}^2)$,

$$\text{where } C_{sb}^2 = \frac{2-\tau_{sb}}{\tau_{sb}} \text{ and } \tau_{sb} = \frac{\tau_s}{\tau_a + (1-\tau_a)\tau_s}.$$

Moreover, the response time of an individual job in a stable GE/GE/1 queue is equal to the response time of a whole bulk in a stable $M(\lambda \tau_a)/GE(\tau_a \mu, C_{sb}^2)/1$ queue (due to geometrically distributed bulk sizes) (c.f.,[18]). However, the response time in a stable $M(\lambda)/GE(\mu, C_s^2)/1$ queue can be shown via spectral analysis (c.f.,[19]) to be also of GE-type, namely, $GE(1-(1-\rho)(1-\tau_s), \tau_s\mu(1-\rho))$, where $\rho = \lambda/\mu$. Thus applying the above result in the case of a stable $M(\lambda \tau_a)/GE(\tau_a \mu, C_{sb}^2)/1$ queue, it follows that the response time distribution for a stable $GE(\lambda, C_a^2)/GE(\mu, C_s^2)/1$ queue of Theorem 1 holds.

Q.E.D.

Appendix II: Proof of Corollary 1:

The response time rv $W \sim GE(\zeta, C_w^2)$ Thus, $P(W \le \pi_w(q)) = q$, $0 < q > 1$, or

$$1 - \tau_w e^{-\zeta \tau_w \pi_w(q)} = q, \text{ or}$$

$$-\zeta \tau_w \pi_w(q) = \ln\left(\frac{1-q}{\tau_w}\right), \text{ or } \pi_w(q) = -\frac{1}{\zeta \tau_w}\ln\left(\frac{1-q}{\tau_w}\right)$$

Q.E.D.

CHAPTER 6

SIMULATION MODELING OF LOCAL AND METROPOLITAN AREA NETWORKS

Marco Conti and Lorenzo Donatiello

6.1. INTRODUCTION

Local Area Networks (*LANs*) are typically based on a high-speed link which is shared among all the stations connected to the network. Information broadcasting is thus easily achieved, and routing is not necessary. In an office building or a university campus LANs represent an efficient and cost-effective way to access servers, to share expensive devices, to exchange electronic mails, etc. With the continuing success of LANs, demand has evolved in the direction of extending their capabilities. The progress in fiber-optic technology, has produced the so-called Metropolitan Area Networks, or *MANs*. MANs represent the evolution of LANs towards higher data rates, e.g., 100-155 Mbps, and coverage up to 100 km.[34,1] On the other hand, the evolution in wireless communications has generated the so-called Wireless LANs (*WLANs*) which extend the LAN services to mobile users.[43,30,8] Hereafter, unless explicitly stated, we use the words LAN and MAN interchangeably.

Since a LAN/MAN network relies on a common transmission medium, Medium Access Control (*MAC*) protocols have been designed to manage the sharing of the bandwidth among the network users.[24,36] The aim of a MAC protocol is to control the interference and competition among users while optimizing overall system efficiency. Performance analysis thus has an important role in MAC protocol selection and design, network tuning, and network-resource allocation.

There are two main approaches in system performance evaluation: the first uses measurement techniques, the second is based on a representation of the behavior of a system via a model. The solution to the model gives the performance indices of the model, which in turn are an estimate of the indices of the system.[17,26,29,31,35,44]

The system performance measurement techniques are applied to a real system, and thus they can be applied only when a real system or a prototype of it is available. Using a model, on the other hand, allows us to study the system in each phase of its life cycle, or rather in each design, development, set-up and modification stage. In this chapter we focus on the use of modeling methods and techniques in LAN/MAN performance evaluation.

Evaluating system performances via models consists of two steps: *i*) defining the system model, and *ii*) solving the model using various solution techniques to get the model performance indices that make up an estimate of the indices of the system. As far as point *i*) is concerned, network performance evaluation models have been defined by using various formal methods such as: queueing theory[33,47] stochastic Petri Nets[31] and stochastic processes.[7] In this chapter we focus on communication system models defined by using queueing network models, *QNMs*.

Models solution methods include analytical and simulative techniques.[17,26,36,38,39] Most of the existing performance studies for high-speed LANs and MANs have been carried out via simulation as it is extremely difficult to analytically solve detailed models of high-speed MAC protocols.[2,5,6,9,10,11,12,13,23]

Throughout this chapter we discuss the use of simulative techniques in the LAN/MAN performance evaluation.[13,24,37]

6.2. LAN/MAN PERFORMANCE MEASURES

The target of a MAC protocol is to share resources efficiently among several users. This efficiency can be expressed in terms of *capacity* and *fairness*.[1,13,36,50] Capacity and fairness are used to evaluate the MAC protocol algorithms, however from the user standpoint other performance figures are needed to measure the Quality of Service (*QoS*) that can be relied on. Below a characterization of these performance measures is given.

CAPACITY. Network stations need some degree of coordination to share the medium of a LAN or a MAN. This coordination is always achieved by means of control information which can be either explicitly carried by control messages traveling along the medium (e.g., reservations, tokens), or implicitly provided by the medium itself (e.g., under the form of channel active or idle). Control messages or message retransmission due to collisions (the latter occurs in MAC protocols which make use of implicit control information) subtract channel bandwidth from that available for successful message transmissions. Therefore, the fraction of channel bandwidth used by successfully transmitted messages gives a good indication of the overhead required by a MAC protocol to perform its coordination task among stations. In the literature the MAC protocol capacity figure, ρ_{max}, is used to characterize this aspect.[1,13,36] This is defined as the maximum fraction of the medium bandwidth used by the nodes, and it is generally measured by assuming that each node tries to

LAN/MAN SIMULATION MODELING

seize all the medium capacity, i.e., *asymptotic conditions*.[10,11]

FAIRNESS. Capacity is a measure of the aggregate bandwidth the network users can rely upon. However, it doesn't say anything about the way this bandwidth is subdivided among the users. Therefore an additional performance measure named *fairness* is introduced. Fairness means that the network does not differentiate between stations in granting them access rights to the transmission bandwidth.[13,50] In this chapter we use the *access delay* (i.e., the delay between packet generation and transmission) as a fairness metric when the network operates under light- and medium-load conditions. On the other hand, when the offered load is greater than the network capacity the *throughput* is the fairness metric. In any case, these fairness indices will be evaluated by assuming that each node generates the same average traffic and hence, in an ideal fair system, each node should experience the same access delay, and throughput.

USER REQUIREMENTS. In *legacy* LAN networks, whose main target is the support of EDP data applications, e.g., file transfer, e-mail, etc., the quality of service is generally expressed in terms of average access delay and throughput; while for integrated services networks (e.g., high-speed LANs, MANs, WLANs) which have the potential to support, beyond EDP data applications, time-constrained applications (e.g., voice and video) as well, the performance figures distribution would also be necessary. In fact, the most important requirement of a time-constrained application is the percentage of packets which are correctly delivered with a delay lower than a given threshold. Therefore, the access delay distribution and the packet-loss probability are the main performance figures used to characterize the application's QoS in a MAN.[24,36]

6.3. LAN/MAN SIMULATION MODELING

The computer based discrete event simulation is one of the most flexible methods (tools) for the performance evaluation of complex systems, such as LAN/MANs. The goal of a simulation study is the construction of a simulator that mimics the system state transitions, and, (by using the data collected during the simulation) the estimation of the performance indices of the systems under analysis. An orthodox simulation study is based on several steps whose characteristics and number can vary with respect to the nature of the system analyzed and to the objectives of the study.[29,38,39,44] The key steps establishing the kernel of any simulation study are: *i) problem formulation*; *ii) workload characterization*; *iii) model definition and validation*; *iv) construction and verification of the simulator*; *v) design of experiments*; *vi) analysis of the simulation results or output analysis*. A careful discussion of the various steps of a simulation study can be found in several books and papers.[18,26,29,32,35,37,38,39,44,45] In the next subsections, first we discuss the problem of the workload characterization, then we focus on the structure of the simulation program which represents the evolution of the network state, and finally we consider the statistical analysis aspects of a simulative experiments. We shall discuss each step by considering as systems of interest local and metropolitan area networks.

6.3.1 Workload Characterization

The performance study of a computer-communication system requires a specification of the users demand for hardware and software resources (*workload characterization*). This may include memory space, CPU processing time, and channels for information transfer.

The quantitative characterization of the workload requires the specification of the time instants at which resource requirements are issued, the type of requested resources, the amount of processing time, etc.

In the evaluation of a LAN/MAN protocol the users are the application programs which cooperate by exchanging messages through the network. The resources are the network bandwidth required to transmit the messages, and the buffer-space to store packets at the sender and at the receiver side. The user service requests correspond to the transmission of the messages. Hence, the workload characterization in a LAN/MAN environment requires the specification for each application of the following parameters:

- the time instants at which the sender application inserts the messages in the queue of the network station to which it is connected;
- the destination address of the receiver. This information is required to characterize the propagation delay between the sender and the receiver. In addition, in a multichannel network (e.g., DQDB, see Section 6.4), the destination address identifies the channel to utilize to transfer the information to the receiver;
- the length in bits of each message. This information characterizes both the amount of buffer space required to store a message in the sender and receiver buffer, and the message transmission time. It is worth pointing out that messages are often partitioned into smaller units, packets, and thus the length of a message is also expressed in number of packets.

Two characterizations of the workload have generally been used in performance modeling studies: synthetic workloads and traces of a real workload.

SYNTHETIC WORKLOADS. The simplest parameter to characterize a workload in a networking environment is the *Offered Load* (*OL*). The offered load is an average figure which designates the aggregate user-generated traffic normalized to the medium capacity. More precisely, the aggregate user-generated traffic is the total number of bits sent, in a time unit, by the applications to the network. The *Workload Type* parameter is used in the literature[9-12] to specify the contribution of each station to the offered load. Average values (e.g., the offered loads) alone are not sufficient if the variability in the workload must be characterized. In this case the *message length* and the *message interarrival time* distributions need to be specified. In computer networks users' messages are segmented (at the sender side) into packets. The message length distribution, $F_{Msg}(x)$, characterizes either the number of bits in a message or the number of packets contained in a message. The message interarrival time distribution, $F_A(x)$, describes the distribution of the length of the interval between the generation of two consecutive messages. Typically, the message interarrival time distribution is approximated by an exponential or a hyperexponential distribution.[11,12] To clarify these concepts it is useful to refer to Figure 6.1 which shows the characterization of the workload generated by a generic application in a LAN/MAN environment.

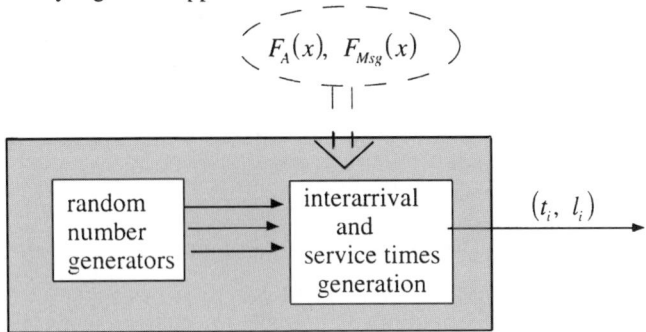

FIGURE 6.1. Synthetic workload characterization.

As shown in Figure 6.1, in the simulator the traffic generated by an application is characterized by the sequence of couples (t_i, l_i) which identifies the time at which a message is inserted into the station queue and the length of this message. The sequence $\{(t_i, l_i), i = 1, 2, ...\}$ is generated by utilizing[38,39]

i) two (or more) random number generators which produce two independent sequences of random numbers sampled from a uniform distribution in the interval (0,1); and,
ii) the methods for generating a sequence of random observations sampled from $F_A(x)$, and $F_{Msg}(x)$ by exploiting the sequences produced by the random number generators.

TRACES. Synthetic workloads have the advantage that they can be made parametric and hence flexible. For example, the workload can be immediately scaled by varying the value of the offered load. Traces are generally used when synthetic workloads are not able to provide an adequate representation of the features of a real application. For example, traces are often used in the analysis of bandwidth allocation problems when *Variable Bit Rate* (*VBR*) video sources are involved.[2] A trace is obtained by recording for each real application the messages generation instants, the destination addresses, and the length of the messages. In this case, as shown in Figure 6.2, the characterization of an application in the simulator is very simple. The sequence $\{(t_i, l_i)\}$ is generated by simply

reading into a file. However, traces are expensive to obtain (measurements of real systems are required) and to manage (large files need to be manipulated). In addition, traces are not flexible:
- a trace which characterizes a given application (e.g., a coded movie) cannot be used to characterize an application of the same class. A trace must be recorded for each application;
- traces have a finite length and this may be not adequate to obtain reliable simulative estimates;
- traces are static and cannot be used in parametric studies in which we wish to scale the parameters of the performance study.

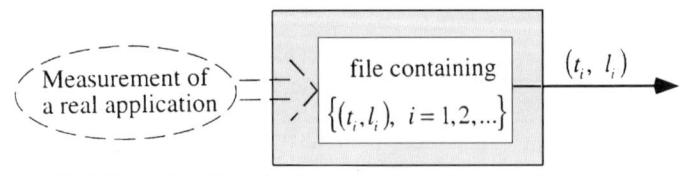

FIGURE 6.2. Workload characterization based on traces.

SYNTHETIC WORKLOADS BASED ON MARKOV PROCESSES. A synthetic workload based on distributions is not able to capture, for example, the temporal dependencies of VBR video sources.[14,15] On the other hand, the use of traces in performance studies has several drawbacks. An alternative approach which reduces the approximation of a synthetic workload characterization, and which is more flexible than real traces, is the synthetic workload characterization based on Markovian models.

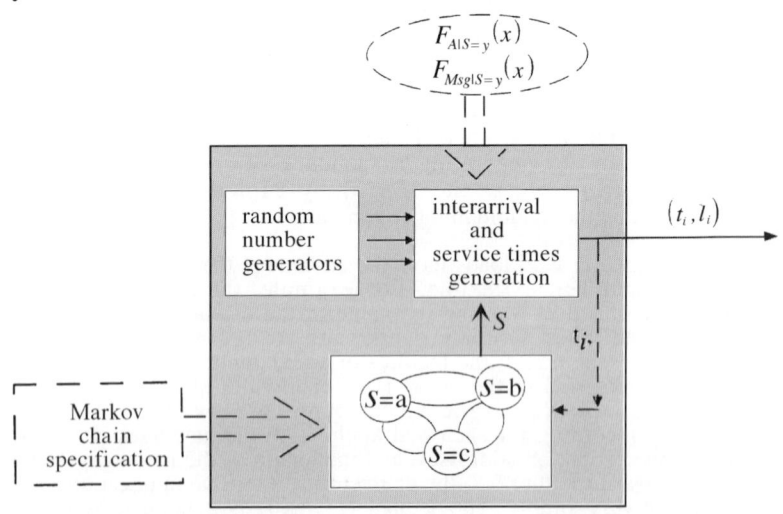

FIGURE 6.3. Synthetic workload based on a Markov model.

As shown in Figure 6.3, a synthetic workload characterization based on a Markov chain generalizes the synthetic workload characterization based on distributions. In this case the interarrival times and the message-length distributions may depend on the state S of a Markov chain. The transition in the Markov chain, generally occurs after each message generation, and for this reason in the figure there is a feedback from the block which characterizes the message generation instant and the Markov chain.

6.3.2 Construction of a Computer Simulation Program

A simulator can be built by using general-purpose languages like Pascal, C or Simula (in spite of its name, Simula is a general purpose language with features to construct simulators), or special-purpose languages like GPSS or SLAM, or in the case of QNMs, software packages like RESQ or QNAP.[18,37,45]

Two approaches can be used for the definition of the simulator's structure, *the event scheduling approach* and *the process interaction approach*.[18,35,38,39] In the event scheduling approach, an event contains all the information required to perform all the system changes that this event causes. A system simulation corresponds to the execution of a (chronologically ordered) sequence of events. The process interaction approach provides a process for each entity in the system. Processes move into the system and sometimes their movement in the system is delayed, for example, when a resource is not available. Here, we shall discuss the event scheduling approach in the context of general purpose languages. The process approach in developing a system simulator is discussed by Franta[19] in the framework of Simula.

The principal components of a simulator based on the event scheduling approach are the following:

a) a set of variables to describe the model state (*state variables*);
b) a variable, *simulation clock*, whose current value denotes the simulated time;
c) an *event list*. Each element in this list contains the event type and the next occurrence time of the event;
d) *event routines*. For each event type, we construct an event routine which implements the state transformation function connected to the specific event. The event routines also update the event list. At the end of the computation, each event routine transfers control to the control routine;
e) *control routine*. Based on the occurrence time of the events contained in the event list, this routine determines the next event to elaborate, updates the simulation clock and transfers control to the selected event routine;
f) a *statistics routine* to collect data which will be used to obtain the estimation of the performance indices.

The procedures for the generation of random variables are important support routines for any simulator. The random nature of interarrival times, service and resources requests is represented by means of routines which generate random numbers characterised by specific probability distributions. The methods for generating random variables are analyzed, for example, by L'Ecuyer.[40,41]

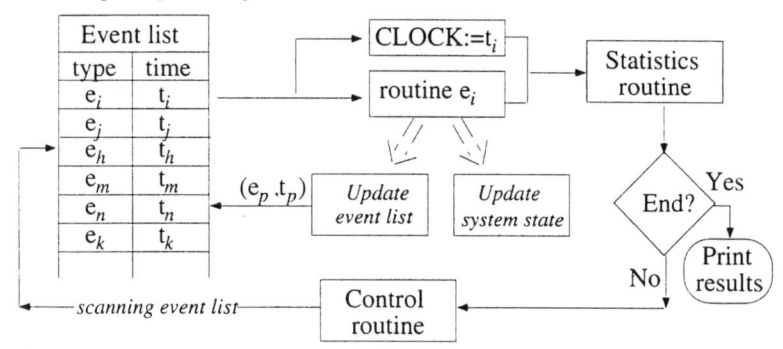

FIGURE 6.4. Asynchronous timing.

The temporal evolution of the model is realized by modifying the simulation clock variable. There are two main methods for managing the advance of the simulated time: *asynchronous* and *synchronous timing*.

With asynchronous timing (see Figure 6.4), the value of the simulation clock variable always coincides with the minimum (most imminent) occurrence time of the events contained in the event list. This mechanism of time advance is know as next-event time advance and is the most efficient in the simulation of a wide class of QNMs.

On the other hand, with synchronous timing, the (simulation) clock is advanced by a fixed amount Δ. All the events which occur during a Δ interval are managed simultaneously at the end of the interval. If many events occur in an interval this may introduce some approximations in the characterization of the real-system behavior. Moreover, too small a Δ interval (i.e., the probability that an event occurs in Δ is low) may produce a slow advance in the simulation time (at the end of an interval, a check is made to verify if one or more events occurred in the interval), and hence it increases the simulation computational-complexity.

We conclude this section by discussing an alternative time advance mechanism in which the advance of the simulation clock is event-driven (as in asynchronous timing), but as in synchronous timing *i*) the clock is advanced by fixed amounts $\delta_i(t)$, and *ii*) several events may occur at the same time. This mechanism is suitable for describing the dynamic evolution of discrete-time queueing networks or queueing networks models of "slotted" systems.[10,11] The characteristic of these models is that more than one event can occur at a specific time instant (simultaneous events). Moreover, for each set of simultaneous events, the occurrence times are spaced out of $n \cdot \Delta$ units of simulated time, where Δ is the slot duration (i.e., the time unit in a slotted system). Based on these characteristics, as shown in Figure 6.5, the simulator can be constructed by partitioning the events set into subsets of simultaneous events

$\{E_1, E_2, ..., E_M\}$ and by using a time advance mechanism which advances the simulation clock by fixed time increments. It is worth pointing out the reductions in the computational complexity that are possible with this timing mechanism:
- the control routine is reduced to the updating of the counter which identifies the next event;
- the event list becomes a "static" structure, there is no need to update it for inserting a new element, nor to search for the next event;
- the control routine is invoked only once every k events, where k is the average number of simultaneous events.

A simulator based on this timing mechanism applied to the study of a DQDB network will be presented in Section 6.4.1.

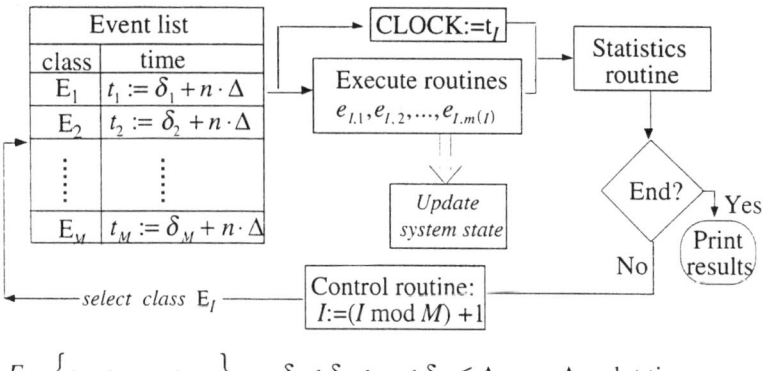

$$E_i = \{e_{i,1}, e_{i,2}, ..., e_{i,m(i)}\} \qquad \delta_1 < \delta_2 < \cdots < \delta_M \leq \Delta \qquad \Delta = \text{slot time}$$

FIGURE 6.5. Simulation clock advancing in slotted systems.

6.3.3 Output Analysis

Two types of simulation can be considered with respect to the output data analysis: terminating simulation and steady-state simulation.[26,29,32,38,39,44] Both types of simulation are used in a LAN simulation study.

TERMINATING SIMULATION. The simulation model is analyzed during a specific and finite interval of simulated time and the performance measures (indices) are calculated with respect to this interval. The length of the interval can be either deterministic or stochastic. This kind of simulation strictly depends on the initial state and can be used to estimate a performance index in a time interval which is critical for the system. For example, telephone networks are analyzed during peck-rate traffic periods (e.g., 9 am, 12 am) to study the blocking probability, i.e., the probability that a call cannot be accepted by the network. Transient simulation is a special case of terminating simulation. In transient simulation the system behavior is observed from an initial state up to the time instant at which it enters in a steady-state condition.

STEADY-STATE SIMULATION. The aim of this kind of simulation is to estimate the quantities of interest of the system in equilibrium conditions, i.e., the system state is represented by a stationary process. In this case the model is analyzed by observing it on an time interval which is ideally infinite.

The objective of a simulation study is the estimation of the characteristics (e.g. mean, variance, percentile of the probability distribution) of a random variable, say X, which represents a performance index of the system under investigation. To perform this task, distinct observations of the variable have to be made during the simulative study and statistical inference have to be performed on the set of observations. Generally, the observations of the random variable collected during a simulation run are correlated (i.e., not independent). Hence, it is not possible to directly apply the classical statistical techniques to the output of a simulation study to infer confidence intervals for the estimated performance indices. Several methods have been proposed either to remove the correlation among observations or to exploit the nature of such a correlation.[18,26,29,32,38,39,44] An additional problem occur in a steady-state simulation. During the initial phase of the simulation (transient phase) the process which describes the state of the system is not stationary, and hence the observations of a random variable collected during this phase are correlated and not identically distributed. To overcome this problem, the data gathered during the transient phase are discarded so that only data collected during the stationary phase (which are identically distributed) are used for statistical analysis.[18,38,39,44]

The method of independent replications is one of the most appealing statistical techniques to obtain point estimations and confidence intervals for the quantities of interest both in terminating and in steady-state simulation. Below, this method is briefly sketched. The statistical analysis of a steady-state simulation output is surveyed by Pawlikowski.[44] The estimation of both transient and steady-state performance figures is discussed by Welch.[38]

INDEPENDENT REPLICATIONS. The method of independent replications circumvents the problem of the autocorrelated nature of the simulative observations, by repeating the simulation several times and each time using a different and independent sequence of random numbers.

In the case of terminating simulation, the characteristics of interest of the random variable X are estimated by performing n replications of the simulation and each replication provide an independent observation of X. By denoting with X_i the value of the random variable X provided by the i-th replication of the simulation, the values $X_1, X_2, ..., X_n$ are used to obtain both the point estimation and the confidence interval for the measure of interest.[38]

In the case of steady-state simulation, once the transient phase has been removed, the observations of the specific random variable made during a run, are used to obtain point estimates of the measures of interest. The independence of the point estimates, obtained from different runs, is guaranteed by the independence of the runs. Finally, the independence of the point estimates allow us to apply standard statistical techniques to calculate a confidence interval for the measure of interest. For example, let us consider the estimate of the mean value of X. The objective of the simulation is to obtain the point estimate and confidence interval for the mean μ of X. Let $x_{i,j}$ be the j-th observation of X in the i-th independent replication of the simulation, $i = 1,2,...,n$, $j = 1,2,...,k$. k is the number of observations made in a replication. Each replication provide a point estimate of the mean

$$\tilde{\mu}_i = \sum_{j=1}^{k} x_{i,j}/k$$

$\tilde{\mu}_i$, $i = 1,2,...,n$ are independent and identically distributed, hence we can obtain an estimate of μ by using the classical estimator

$$\hat{\mu} = \sum_{i=1}^{n} \tilde{\mu}_i/n.$$

To generate a confidence interval for μ, we have to calculate the sample variance, $\sigma^2(\hat{\mu})$,

$$\sigma^2(\hat{\mu}) = \sum_{i=1}^{n} (\tilde{\mu}_i - \hat{\mu})^2 / (n-1) .$$

The random variable

$$\frac{\hat{\mu} - \mu}{\sigma(\hat{\mu})/\sqrt{n}}$$

has, approximately, a t-distribution with $n-1$ degrees of freedom.[35,38] By denoting with $t_R(y)$ the y-th percentile of the t-distribution with R degrees of freedom, we have

$$P\{t_{n-1}(\alpha/2) \leq \frac{\hat{\mu} - \mu}{\sigma(\hat{\mu})/\sqrt{n}} \leq t_{n-1}(1 - \alpha/2)\} \approx 1 - \alpha$$

where $1 - \alpha$ represents the confidence level for the confidence interval, i.e., the probability that the confidence interval contains the performance measure of interest. Specifically, based on the symmetry of the t-distribution, i.e. $t_{n-1}(\alpha/2) = -t_{n-1}(1 - \alpha/2)$, it can be verified that μ belongs, with probability $1 - \alpha$ to the interval

$$\hat{\mu} \pm t_{n-1}(1 - \alpha/2) \cdot \sigma(\hat{\mu})/\sqrt{n}.$$

Similarly, it is possible to determine confidence intervals of different measures of interest, such as the variance of X, and the percentiles of its probability distribution.[38] To conclude this section, same considerations on the confidence interval width are necessary. Since the value of the function $t_{n-1}(1 - \alpha/2)$ becomes approximately constant for $n \geq 10$ the confidence interval width is proportional to the ratio $\sigma(\hat{\mu})/n^{1/2}$. To reduce the width of the confidence interval we can increase either the number of replications, n, or the number of observations, k, collect during each replication. The presence of the transient phase in each simulation run, suggests to keep n small, say 10-15, and k large so we can minimize the number of data to be discarded due the presence of the transient phase.[35]

6.4. MAC PROTOCOLS DESIGN

One of the most well-known examples of the use of simulative techniques in the MAC protocol design phase is represented by the definition of the Distributed Queue Dual Bus (*DQDB*) protocol.[28] DQDB is the IEEE802.6 standard for a subnetwork of a MAN.

Before presenting the use of simulative techniques in DQDB protocol design it is useful to present the basic features of the protocol.

6.4.1 The DQDB Protocol

The basic structure of a DQDB network is shown in Figure 6.6. The network consists of two high speed unidirectional buses carrying information in opposite directions. The network nodes[#] are distributed along the two buses and they can transmit information to and receive information from both buses, as shown in Figure 6.6. The node in the leading edge of each bus is designated as the *head of* its corresponding *bus* (*HOB*). Each HOB continuously generates slots of a fixed length (53 octets) which propagate along their respective buses. The first byte in a slot constitutes the *Access Control Field* (*ACF*), which is utilized by the nodes in the network to co-ordinate their transmissions.

Without loss of generality, we name bus A the *forward* bus, and bus B the *reverse* bus. In each node, the segments, on arrival, are put in the correct *local node queue* (*LQ*), as determined by the destination address (there are two local node queues, one for each bus). Below we will focus on segment transmission by using the slots in the forward bus, since the procedure for transmission in the reverse bus is the same.

To manage the slot access mode, the ACF includes a *busy bit* and a *request* (*REQ*) *bit*. These bits are set to "0" by the originating HOB. The busy bit indicates whether or not the corresponding slot has already been used for data transmission. The procedure for segment transmission on the forward bus utilizes the busy bits in the ACF of the slots of the forward bus, and the request bit (the REQ bit below) in the reverse bus. Each node is either *idle*, when there is nothing to transmit, or *count_down*. When it is idle the node keeps count, via the *request counter* (*RQ_CTR*), of the number of outstanding REQs from its downstream nodes. The RQ_CTR increases by one for each REQ received in the reverse bus and decreases by one for each empty slot in the forward bus. When a node in the idle state receives a segment, it enters the count_down state and starts the transmission procedure by taking the following actions: 1) the node transfers the contents of the RQ_CTR to a second counter named the *count_down counter* (*CD_CTR*), 2) resets the RQ_CTR to zero, and 3) generates a request which is inserted into the queue of the pending requests while waiting for transmission on the reverse bus (by setting REQ=1 in the first slot with REQ=0). In the count_down state the CD_CTR is decreased by one for every empty slot in the forward bus until

---

[#] We will use the terms node and station interchangeably.

LAN/MAN SIMULATION MODELING

it reaches zero. Immediately afterwards, the node transmits the segment into the first empty slot of the forward bus. In the meantime, the RQ_CTR increases by one for each new REQ received in the reverse bus from the downstream nodes. After the segment transmission, if the LQ is empty the node returns to the idle state, if not the transmission procedure (1-3) is repeated.

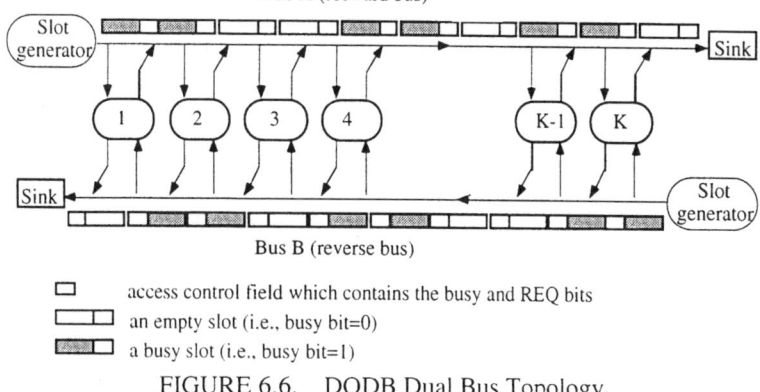

FIGURE 6.6. DQDB Dual Bus Topology.

6.4.2 The DQDB Protocol Evaluation

The importance of DQDB has meant that its performance has been extensively analyzed during its standardization process. Most of the existing results have been obtained via simulation. Specifically, while it can be easily verified that the protocol capacity is equal to one (i.e., $\rho_{max} = 1$) the analysis of fairness and user-oriented performance figures requires the use of simulative techniques.[11,13]

Depending on the amount of traffic the network stations transmit, DQDB simulative analyses have been carried out both in *normal conditions* (i.e., when the offered load is lower *[underload]* or slightly higher *[overload]* than the medium capacity), and in *asymptotic conditions* (i.e., when each DQDB MAN node is trying to seize all the medium capacity). In normal conditions the network model must take into account the randomness of user demands, and this leads to a *stochastic simulation* which obviously contains problems related to the statistical reliability of the results (see Section 6.3.3). By contrast, in asymptotic conditions, the network behavior is often completely deterministic, and this leads to a *deterministic simulation*.[32]

Simulative results related to the DQDB behavior in normal and asymptotic conditions are presented below. The simulative results, obtained with the independent replication technique and a confidence level equal to 90%, refer to a 50 node network. The network length is about 90 km and nodes are equally spaced.

Before presenting the simulative results of DQDB it is worth pointing out some of the features of the DQDB behavior which facilitate its simulation.

DQDB SIMULATION. As observed in Section 6.3.2, for slotted systems, system simulation can be made more efficient by taking into account that several events occur at the same time instant. In the DQDB protocol, the state transitions are triggered by (the header of) the slots traveling on the two buses, hence, in the simulation of a DQDB network, a significant computational-complexity reduction can be achieved by assuming that the distance between two adjacent nodes is equal to an integer multiple of the slot length* Δ. In this case, in fact, all nodes observe the slots on the forward (reverse) bus at the same time instants. More precisely, by denoting with δ the temporal shift among the two buses (i.e., the interval which elapses from the time instant at which a node observes the start of a slot on the forward bus and the arrival of a slot on the reverse bus), the simulator event list always contains only two events related to the arrival of a slot on the forward and reverse buses, respectively. The latter event occurs δ time units after the former (see Figure 6.7).

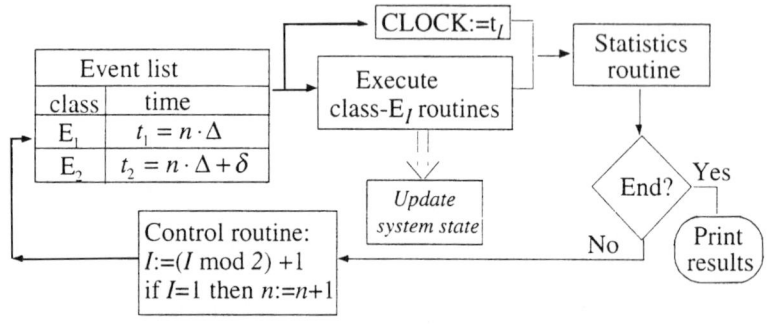

FIGURE 6.7. Timing in a DQDB simulator.

The data structures required to implement such a simulator include for each bus: *i*) an integer array of size M where each element of the array represents a slot traveling on that bus; and *ii*) an integer variable, $First\_x$ ($x \in \{A, B\}$) which denotes the slot which is currently observed by the HOB station on bus x.[#] In addition, for each node a list is required containing the packets queued in this node and two integer variables, CD and RQ, which represent the value of its CD_CTR and RQ_CTR counters, respectively.[28]

---

* By one slot length, we mean the distance between the first and last bit of a slot on the medium.
[#] The arrays are assumed to be cyclic, i.e., given that a station i observed the slot of index j at time t_o, the same station, at time $t_o + \Delta$, will observe the slot of index j-1 if (j-1)>0, otherwise it will observe the slot M.

DQDB FAIRNESS ANALYSIS. In underload conditions, DQDB behavior is satisfactory even though it is not fair.[11,12] The DQDB algorithm only causes a difference in the (bus) access delay experienced by a node, but these access delays are always less than a few milliseconds.[12] In overload and asymptotic conditions, DQDB behavior is unpredictable.[9-12,49] As shown by the simulative results presented below, in these conditions packet loss and throughput depend on the state in which the system operates when the congestion condition occurs. Specifically, when the offered load exceeds the network capacity the DQDB MAC protocol behavior very much depends on the *scenario*, i.e., node activation temporal sequence.[#, 9-12] DQDB behaviors have been analyzed in the literature under several scenarios, e.g., *forward* (i.e., activation is performed from node index 1 to K), *backward* (i.e., activation is performed from node index K to 1) and *random*.[11] The worst case was achieved in the forward activation scenario. The results presented in this chapter are related to this scenario.

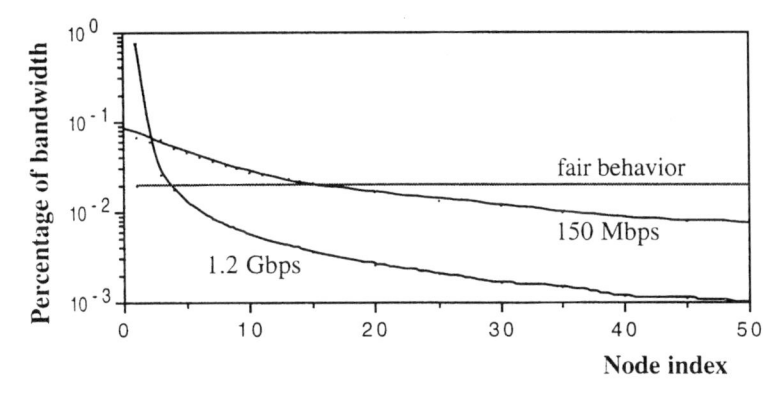

FIGURE 6.8. Throughput on bus A in limiting conditions at 150 Mbps.

Figure 6.8, which plots the bandwidth sharing among nodes for different speeds of the transmission medium, highlights the DQDB unfairness. To understand the dependency of this unfairness on the medium capacity, the curves related to 150 Mbps and 1.2 Gbps must be compared. These curves highlight an amplification in the difference between the percentage of bandwidth of the most and least favoured nodes as the medium capacity passes from 150 Mbps to 1.2 Gbps. This amplification was expected. In fact, DQDB bandwidth sharing depends on the information carried by the slots on the reverse bus (REQ). The number of slots on each bus at 1.2 Gbps is about ten times higher than that at 150 Mbps. This causes an increase in information which has already been sent but not yet received by all nodes (i.e., REQs traveling on the reverse bus).

A scenario defines the order in which nodes become active. When a node is active it operates in asymptotic conditions.

6.4.3 DQDB Protocol and the BWB Mechanism

DQDB unfairness in saturated conditions arises because the protocol enables a node to use every *unused slot* (i.e., an empty slot not reserved by downstream nodes) on the bus. To overcome this problem the *Bandwidth Balancing (BWB)* mechanism[25] has been added to the DQDB standard.

The Bandwidth Balancing mechanism follows the basic DQDB protocol, except that a node can only take a fraction of the unused slots. Specifically, the Bandwidth Balancing mechanism limits the throughput of each node by using a counter, named BWB_CTR, to keep track of the number of transmitted segments. Once the counter reaches the value of a protocol parameter, named BWB_MOD, this counter is cleared and the RQ_CTR is increased by one. This "dummy" REQ limits the throughput a node can achieve.[25] The value of BWB_MOD can vary from 0 to 63. The value 0 means that the BWB mechanism is disabled.[28]

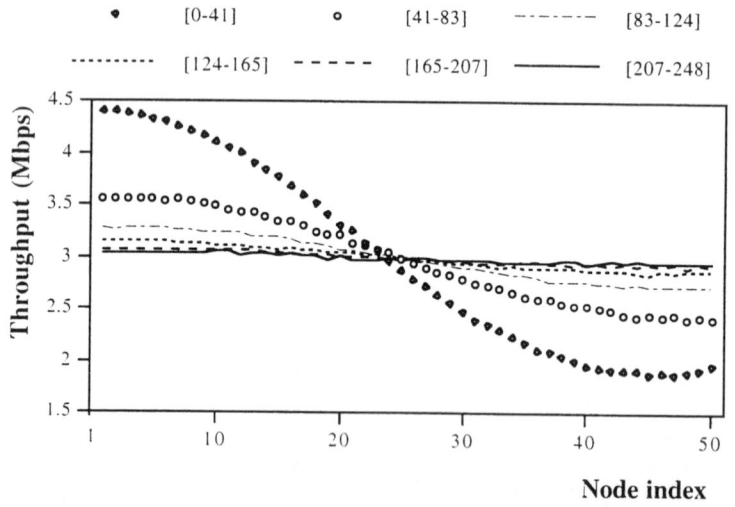

FIGURE 6.9. Effectiveness of the BWB mechanism (BWB_MOD=8).

The effectiveness of this mechanism has been extensively studied via simulation.[5,9-12,49] By considering the same scenario analyzed in Figure 6.8 (e.g., forward scenario and asymptotic conditions), the simulative results presented in Figure 6.9[#] show that DQDB with the BWB mechanism enabled, reaches (after a transient time) a steady-state condition where the bandwidth is equally shared among the nodes. During the transient time the network behavior remains unpredictable.

[#] Curves represent the nodes throughput in consecutive time intervals (milliseconds).

LAN/MAN SIMULATION MODELING

The DQDB behavior presented in Figure 6.9 has also been observed for different BWB_MOD values, network speeds, and scenarios. In particular, the length of the transient interval depends significantly on the BWB_MOD value[#] and initial state, while it is not significantly affected by medium capacity and bus length.[10,11]

The asymptotic conditions should never occur in a real environment, in the next subsection we analyze DQDB behavior under the more realistic *overload conditions*:[9] *i*) the total offered load (for each bus) is slightly greater than the capacity of each bus, and *ii*) the buffer capacity of each node is limited.

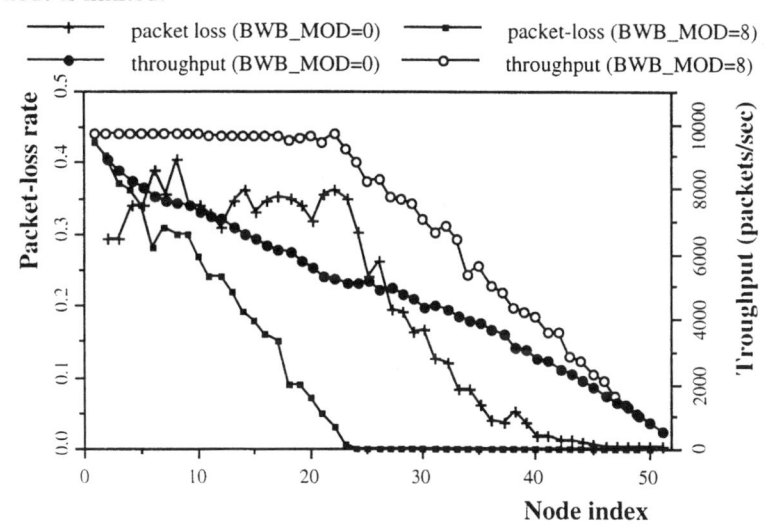

FIGURE 6.10. Throughput and packet-loss (OL=1.20).

6.4.4 Overload Analysis

The aim of this section is to study the degradation of the quality of service experienced by a user when an overload condition occurs. We first analyze the steady state behavior and then the transient behavior.

STEADY STATE ANALYSIS. In overload conditions some nodes achieve a throughput that is lower than their offered load. Thus, for these nodes, the congestion of the local node queue increases greatly and this generates two effects: first, some segments are lost, and secondly, the average access delay almost linearly depends on the local node queue capacity. In

---

[#] In a 50 station network the transient time varies from about 41 msec. when BWB_MOD=1, to 500 msec when BWB_MOD=16.

particular, Figure 6.10 reports the packet loss and the throughput vs. node index in a network with $OL = 1.20$, symmetric workload type[*], single-packet messages and interarrival times exponentially distributed. The figure shows that there are marked differences depending on the BWB_MOD value. In fact, when BWB_MOD=8 (BWB mechanism enabled) nodes with an offered load lower than the minimum bandwidth guaranteed by a fair network (i.e., $(1/K) \cdot \rho_{max}$ which corresponds to 7057 packets/sec in our experiment) do not experience packet loss.[#] On the other hand, with BWB disabled packet loss also affects the stations with an offered load lower than the minimum guaranteed bandwidth.

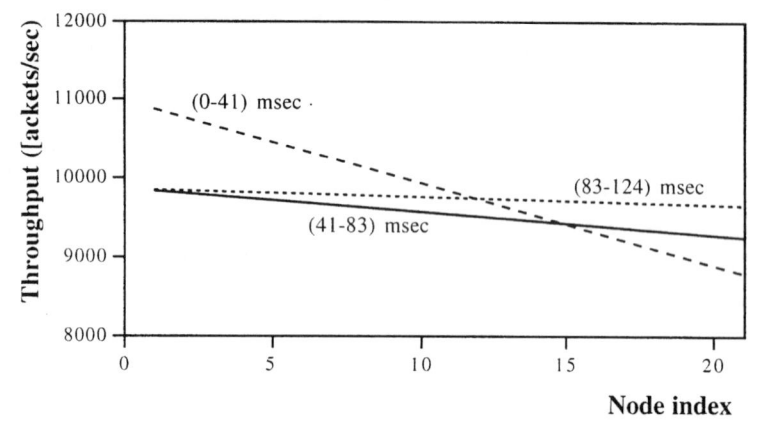

FIGURE 6.11. Throughput achieved by congested nodes.

TRANSIENT BEHAVIOR. In the previous section we showed that, when BWB is disabled, DQDB behaves unpredictably. Thus, if the offered load becomes higher than the medium capacity, the BWB is necessary in order to guarantee (in steady-state conditions) a minimum quality of service to all nodes. However, Figure 6.9 (BWB_MOD=8) showed that, when congestion occurs, it may take a "long" transient period to reach the steady-state condition. Figure 6.11 shows DQDB transient behavior when the BWB is enabled with a symmetric workload type and OL=1.20. The steady-state is reached after about 120 milliseconds, and thus the transient

---

[*] In a symmetric workload type all nodes generate the same traffic and the destination address is uniformly distributed, i.e., the probability that a node transmits a segment to any other node is constant. As a consequence, in DQDB, the probability that a node transmits a segment on a given bus is proportional to the number of downstream nodes on that bus.

[#] With BWB enabled, $1/K$ is a lower bound of the guaranteed throughput on a single bus. This bound is achieved when all nodes are congested. In Figure 6.10 only the first 22 nodes are congested. Thus they achieve a much higher throughput (about 9600 packets/sec) than the minimum guaranteed.

time is not negligible at all.

6.5. BANDWIDTH ALLOCATION

Among the prominent features of high-speed networks is service integration. The network provides a low-cost packet transport service attracting customers with many different types of traffic. Voice users require an upper bound in delay with which information is transferred from source to destination. On the other hand, they can tolerate a certain degree of delay variability and packet loss. Video users can tolerate a larger end-to-end delay but they are sensitive to delay variability and packet loss. The transmission of an alarm message needs both a delay constraint and high reliability. Data applications require only a reliable transportation service. By suitably mixing the access to the network by different sources, a MAN technology must ensure sufficiently high utilization of the medium to justify its cost, while still guaranteeing the QoS required by each type of traffic. Specifically, the network bandwidth must be guaranteed to the real-time applications, while the residual bandwidth may be utilized for non real time traffic. Furthermore, to maximize the number of real-time applications that can concurrently use the network, the minimum amount of bandwidth that must be reserved for each application (to satisfy its QoS requirements) must be identified.

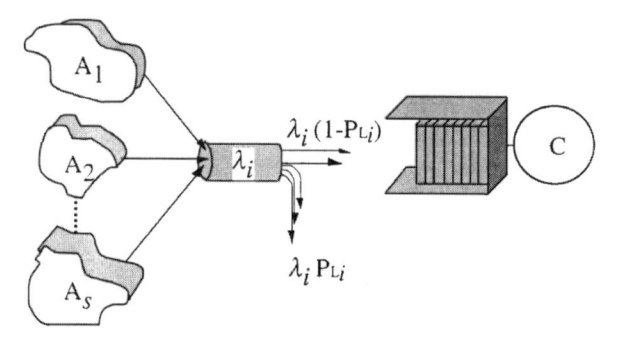

FIGURE 6.12. Statistical multiplexing of VBR video sources at node $\{i\}$.

Variable Bit Rate (*VBR*) video is by far the most interesting and challenging real-time application. A variable bit rate encoder attempts to keep the quality of video output constant at the price of changing the bit rate. A better utilization of network resources is also obtained since only the effective amount of information has to be transferred. However, this makes the bandwidth allocation for this traffic very difficult since: *i*) VBR video traffic is highly variable and dependent on the adopted coding scheme and movie activity, *ii*) these applications have low tolerance towards network congestion as information losses may result in severe degradation in the information at the receiver side; and *iii*) although sufficient buffer capacity may be available at the sender side to absorb congestion, excessive buffering may not be possible, due to the resulting unacceptable delays. A conservative bandwidth allocation approach is often used to avoid the above problems: the peak-rate bandwidth is reserved for each VBR video application. However, since the ratio peak/average for VBR video traffic is often greater than five, the peak rate approach significantly reduces the number of VBR applications that can concurrently use the network. To avoid this inefficient approach a careful analysis of the relationship between the QoS and the amount of bandwidth reserved for a VBR video application must be performed. Specifically, in this section we investigate the queueing time distribution experienced by VBR video traffic as a function of the bandwidth reserved for each source. The potential gains (in bandwidth allocation) which can be achieved by multiplexing several i.i.d. MPEG video sources[42] are also investigated. To this end a simulative study is used to estimate the queueing delay distribution in a single server queueing system with deterministic service time, FIFO service discipline and input traffic generated by s independent MPEG-1 sources (see Figure 6.12). Packets are removed from the queue by a server with a rate C which corresponds to the total bandwidth reserved for the s sources. b denotes the amount of bandwidth reserved for each source, hence $C = s \cdot b$.

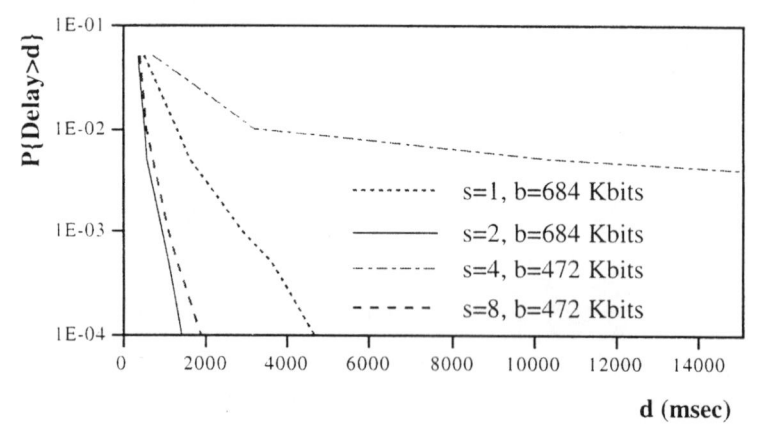

FIGURE 6.13. Statistical multiplexing of VBR video sources.

As the traffic generated by an MPEG-1 encoder has both a high degree of burstiness (peak/average ratio greater than five) and a strong long term correlation, in the simulative study, the MPEG sources have been characterized by using a synthetic workload based on a Markov chain (see Section 6.3.1). Specifically, a 64 states bidimensional Markov chain was utilized by fitting the chain parameters to the MPEG-1 trace of the "Star Wars" movie.[14,15]

LAN/MAN SIMULATION MODELING

Taking into account that the "Star Wars" movie produces a trace with an average bit rate equal to 374 Kbits/sec the QoS experienced by the applications for two different bandwidth allocations was analyzed: b=472 and b=684 Kbits/sec per source, respectively. Figure 6.13 indicates that for a reserved bandwidth equal to (about) twice the average bit rate per source (i.e., b=684), a significant multiplexing gain is achieved even with only two sources. To further increase network utilization, the number of multiplexed sources quickly increases. For example, as shown in Figure 6.13, 80% network utilization and acceptable delays can be achieved if at least eight sources are multiplexed.

6.6. WIRELESS LANS DIMENSIONING AND TUNING

Wireless LANs (*WLANs*) consist of one base station, several mobile terminals (with a radio interface) and two communication channels: a mobile-to-base-station channel (*up-link*) and a base-station-to-mobile channel (*down-link*).[43] We assume that the up-link and down-link communications are physically separate (e.g., on different frequency channels) and slotted. Fixed-length packets arriving at the mobiles are buffered at the terminals until they are transmitted on the up-link to the base station. The base station either broadcasts packets (addressed to the mobiles within its cell) on the down-link, or transfers the packets to the wired networks. Hence, while the base station is in complete control of the down-link, the up-link is shared among several mobile terminals. As the communication bandwidth in the up-link is limited, the MAC protocol is the main element in determining the network efficiency.

Speech traffic is likely to be the predominant traffic for future wireless networks, hence, WLAN protocols are designed to maximize the number of voice communications which can be active simultaneously inside the network. Furthermore, as voice connections require that the subnetwork guarantees the absence of congestion, bounds on packet transfer delay, and bounds on the packet-loss rate, the MAC protocols limit the number of voice connections inside the network. To identify the maximum number of voice connections which can be simultaneously supported by a WLAN simulative studies are generally used.[8,22,23,30] In this section we focus on the use of simulative studies in the tuning and dimensioning of a wireless network based on the *Packet Reservation Multiple Access* (*PRMA*) protocol.[22]

PRMA ANALYSIS. PRMA is a TDMA protocol[24] in which the terminal-base-station channel is subdivided into fixed-length time slots, and each slot can be used to transmit either a packet of speech or data. In addition, N_s time slots are grouped to form a frame. Each frame is arranged to have the same duration as the voice packetization delay. Hence, a voice terminal requires exactly one slot in each frame. Throughout we assume that only voice terminals are present in a cell. Furthermore, we assume that the suppression of the silent periods is implemented in the voice terminals, i.e., they alternate talkspurts and silent periods and speech packets are only generated during talkspurts.[22] In this case, a synthetic workload characterization based on Markovian models can be used to model the traffic generated by the voice sources (see Section 6.3.1). Specifically the model for voice sources is provided by a two state Markov process: exponentially distributed talking periods alternate with exponentially distributed silent periods.

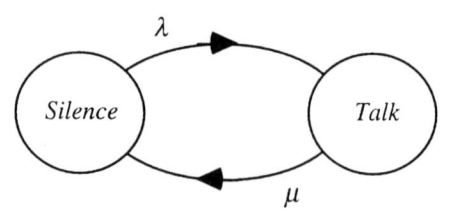

FIGURE 6.14. The voice model.

A Markovian model for a voice source with transition rate λ (μ) from silence (talk) to talk (silence) is given in Figure 6.14. Specifically, the average lengths of a talkspurt and silence periods are $1/\lambda = 1$ sec, and $1/\mu = 1.35$ sec, respectively.[22]

To transmit its packets a terminal must reserve one slot per frame. Slots can be in two states: reserved or unreserved. A terminal, as soon as it generates the first packet of a talkspurt, begins to contend with other terminals for each unreserved time slot, according to the R-ALOHA protocol.[48] During a talkspurt, a new packet is generated at each frame. To successfully reserve a slot, three favourable conditions must be met: the current slot is unreserved, the terminal has permission to transmit, and no collision occurs with packets of other contending terminals. Permission occurs with a fixed probability, p, at each time slot, independently at each terminal. Collision occurs when two or more terminals try to transmit in the same unreserved slot. Upon collision, no packet is transmitted and all colliding packets must be retransmitted. Upon a successful transmission of a speech packet, the terminal obtains a reservation for that slot in the future frames. The reservation mechanism guarantees that all subsequent packets will suffer no collision with packets from other terminals. At the end of the talkspurt, the terminal releases the reservation by leaving the slot empty. While a terminal, in the talkspurt state, is waiting to obtain a reserved slot it discards its generated packets as soon as their holding times exceed the delay limit. As voice connections require that the subnetwork guarantees the correct delivery of about 99 per cent of the generated packets,[22] for a careful network tuning it is important to identify the maximum number of voice terminals that the PRMA protocol can support. To this end extensive simulative experiments have been performed.[22] Several variables influence the PRMA efficiency, such as the value of the permission probability, the number of active terminals, the maximum delay before discarding the voice packets, etc. An extensive simulative study of the impact of all these parameters on the PRMA performance was carried out by Goodman and Wei.[23] In the following, we report the results of a PRMA simulative study obtained by assuming that *i*) the up-link and down-link channel speed is equal to 720 Kbit/sec; *ii*) the up-link is divided into fixed-length frames of length $F = 16$msec; *iii*) each frame is subdivided into 21 slots; *iv*) 32 Kbit/sec speech coding; and *v*) permission probability $p = 0.3$. Table 6.1 reports the packet dropping probability as a function of the number of calls in progress in a cell by assuming that packets are discarded after 32 msec. As shown in the table, the PRMA simulative analysis indicates that the protocol is able to support up to 31 calls, i.e., about 1.5 voice communications per time slot.

TABLE 6.1. Packet-dropping probability in a PRMA network.

Number of voice terminals	Packet-dropping probability	Confidence Interval (C.I.)	C.I. width
24	0.00063	0.00063 ∓ 0.0001	32.4%
25	0.00086	0.00086 ∓ 0.0001	20.8%
26	0.00139	0.00139 ∓ 0.00019	25.0%
27	0.00220	0.00220 ∓ 0.00019	16.0%
28	0.00296	0.00296 ∓ 0.00021	14.5%
29	0.00463	0.00463 ∓ 0.00056	24.2%
30	0.0070	0.0070 ∓ 0.0008	22.8%
31	0.01045	0.01045 ∓ 0.00085	16.5%

6.7. SIMULATION COMPLEXITY

In this chapter we have presented the use of discrete event simulation techniques in the performance analysis of high-speed LAN/MANs. Little attention has been paid to the computational complexity of a simulation experiment. Since the complexity, dimension, and speed of the networks increase or rare events need to be estimated, the simulation of many computer-network problems becomes rapidly intractable (even with the continuing increase in computer speed) with the classical sequential approach described in Section 6.3.2. For example, in a high-speed network, such as DQDB, up to 700,000 packets per second are delivered by the network, so that simulating network behavior for a few minutes is a complex computational task. To overcome these computational problems, advanced simulative techniques, such as *Parallel and Distributed Simulation (PADS)*, *Hybrid Simulation*, and *Fast Simulation of Rare Events* have been developed.

PARALLEL AND DISTRIBUTED SIMULATION. PADS deals with the execution of a simulation program on a parallel computer either by decomposing the simulation program into a set of concurrently executing processes, or running several copies of the same program on several processing elements. A sequential discrete event simulator (see Section 6.3.2), is based on a "main loop" which repeatedly removes the most recent event from the event list, and then processes that event. It is fundamental that the event to be processed is always the most recent, otherwise a future event may affect the past history of the system. Fujimoto[21] and Ayani[3] surveyed existing approaches for solving the synchronization problems generated by executing discrete event simulation programs on a parallel computer. They analyzed the merits and drawbacks of the various techniques.

HYBRID SIMULATION. This technique is based on a *Hierarchical Modeling* (*HM*) of the system to analyze. The solution is based on the decomposition and aggregation method.[16,31] Specifically, starting from the queueing model that describes the system behavior, HM works in two main steps which can be applied recursively. In the first step the model is decomposed into submodels, which are solved in isolation. From this analysis it is possible to define a simpler representation of each submodel. Specifically, in the second step, in the model which describes the system behavior each submodel is replaced with a single server queue with appropriate parameters, and then the *aggregate model* is analyzed by using analytical or simulative techniques. In a large number of cases the Hybrid Simulation technique results as the most efficient technique in performing the two steps of hierarchical modeling. Generally, with Hybrid Simulation (*HS*), each submodel is solved with analytic techniques and then the reduced system model is analyzed with simulative techniques.[4,20] The HS potential speed-up results from the fact that the submodel needs to be solved once (for each possible customer population) to characterize its aggregate representation. Then in the simulation of the entire model, the potentially large number of events that would have occurred when a customer passed through the submodel are replaced by only two events: the arrival and departure of the customer from the composite queue.

FAST SIMULATION OF RARE EVENTS. These are efficient techniques for estimating, via simulation, the probabilities of certain rare events such as long waiting times or buffer overflows. As shown by Heidelberger,[27] the lower the probability of the event to estimate, the larger the sample size must be. For example, to estimate, with classical simulative techniques (see Section 6.3.3) the confidence interval of an event whose probability is 10E-9, with a 99% confidence level and a confidence interval with a 10% width, a sample size of 6.64 x 10E11 is required. The need to estimate rare events frequently occurs in network dimensioning problems (e.g., buffers size). The general approach to speeding up such simulations is to accelerate the occurrence of the rare events by using importance sampling. In importance sampling the system is simulated using a new set of input probability distributions, and unbiased estimates are recovered by multiplying the simulation output by a likelihood ratio. In practice, this means that the required run lengths can be reduced by many orders of magnitude, compared to standard simulations.

6.8. REFERENCES

1. B. W. ABEYSUNDARA, A. E. KAMAL, ACM Computing Surveys, **23**, 221 (1991).
2. G. ANASTASI, M. CONTI, E. GREGORI, L. LENZINI, Computer Communications, **16**, 39 (1993).
3. R. AYANI, "Parallel Simulation", in Lecture Notes in Computer Science, edited by L. Donatiello and R. Nelson (Springer Verlag, Berlin, 1993) **729**, 1.
4. S. BALSAMO, M. CAPPUCCIO, L. DONATIELLO, R. MIRANDOLA, "Some Remarks on Hybrid Simulation Methodology", in Proc. SCS 90, (Calgary, Canada, 1990) 30-37.
5. C. BISDIKIAN, Computer Networks and ISDN Systems, **24**, 367 (1992).
6. I. CIDON, Y. OFEK, IEEE Trans. Commun., **41**, 110 (1993).
7. E. CINLAR, Introduction to Stochastic Processes, (Prentice-Hall, Inc., Englewood Cliffs, N.J., 1975).
8. A.C. CLEARY, M. PATERAKIS, Int. J. Wireless and Information Networks, **2**, 1 (1995).
9. M. CONTI, E. GREGORI, L. LENZINI, "DQDB Under Heavy Load: Performance Evaluation and Unfairness Analysis", in Proc. INFOCOM '90, (San Francisco, CA, 1990) 313-320.
10. M. CONTI, E. GREGORI, L. LENZINI, "Asymptotic Analysis of DQDB", in Proc. EFOC/LAN90, (Munich, FRG, 1990) 259-269.
11. M. CONTI, E. GREGORI, L. LENZINI, IEEE JSAC, **9**, 76 (1991).
12. M. CONTI, E. GREGORI, L. LENZINI, European Transactions on Telecommunications and Related Technologies **2**, 403 (1991).
13. M. CONTI, E. GREGORI, L. LENZINI, "Metropolitan Area Networks (MANs): Protocols, Modeling and Performance Evaluation", in Lecture Notes in Computer Science, edited by L. Donatiello and R. Nelson (Springer Verlag, Berlin, 1993), **729**, 81.
14. M. CONTI, E. GREGORI, A. LARSSON. "On the Relevance of Long Term Correlation in MPEG-1 Video Traffic", in Proc. Broadband Communications '96, (Montreal, Canada, 1996).
15. M. CONTI, E. GREGORI, "Validation and Tuning of an MPEG-1 Video Model", in Performance Modelling and Evaluation of ATM Networks, edited by D. Kouvatsos (Chapman& Hall publishing Company, 1996).
16. P.J. COURTOIS, Decomposability: Queueing and Computer System Applications (Academic Press, New York, 1977).
17. D. FERRARI, G. SERAZZI, A. ZEIGNER, Measurement and Tuning of Computer Systems (Prentice Hall, Englewood Cliffs, NJ, 1983).
18. G.S. FISHMAN, Principles of Discrete Event Simulation, (John Wiley, New York, 1978).
19. W.R. FRANTA, The Process View of Simulation. (Elsevier North-Holland Inc., New York, 1977).
20. V. S. FROST, W.W. LARUE, K.S SHANMUGAN, IEEE JSAC, **6**, 146 (1988).
21. R.M. FUJIMOTO, CACM, **33**, 30 (1990).
22. D. GOODMAN, R. VALENZUELA, K. GAYLIARD AND B. RAMAMURTHI, IEEE Trans. Commun., **37**, 885 (1989).

23. D. GOODMAN, S.X. WEI, IEEE Trans. Veh. Technol., **40**, 170 (1991).
24. J.L. HAMMOND, P.J.P. O'REILLY, Performance Analysis of Local Computer Networks (Addison-Wedsley, 1988).
25. E.L. HAHNE, N.F. MAXEMCHUCK, A.K. CHOUDHURY, IEEE Trans. Commun., **40**, 1192 (1992).
26. P. HEIDELBERG, S.S. LAVENBERG, IEEE Trans. Comput., **33**, 1195 (1984).
27. P. HEIDELBERG, "Fast Simulation of Rare Events in Queueing and Reliability Models", in Lecture Notes in Computer Science, edited by L. Donatiello and R. Nelson (Springer Verlag, Berlin, 1993), **729**, 165.
28. IEEE, Distributed Queue Dual Bus (DQDB) Metropolitan Area Network, (IEEE 802.6, New York, 1990).
29. R. JAIN, The Art of Computer Systems Performance Evaluation (Wiley, New York, 1991).
30. D.G. JEONG, C.H. CHOI, W.S. JEON, IEEE/ACM Trans. Networking, **3**, 742 (1995).
31. K. KANT, Introduction to Computer System Performance Evaluation, (McGraw Hill Int. Editions 1992).
32. J. P.C. KLEIJNEN, Statistical Techniques in Simulation, **1**, (Dekker INC., 1974).
33. L. KLEINROCK, Queueing Systems,1 (Wiley, New York, 1975).
34. R. W. KLESSING, IEEE Commun. Mag., **24**, 9 (1986).
35. H. KOBAYASCHI, , Modeling and Analysis, (Addison Wesley, Reading. MA, 1978).
36. J. F. KUROSE, M. SCHWARTZ, Y. YEMINI, ACM Computing Surveys, **16**, 43 (1984).
37. J.F. KUROSE, H.T. MOUFTAH, IEEE JSAC, **6**, 130 (1988).
38. S.S. LAVENBERG, Computer Performance Handbook, (Academic Press. New York, 1983).
39. A.M. LAW, W.D. KELTON, Simulation Modeling and Analysis, (McGraw Hill, New York, 1982).
40. P. L'ECUYER, CACM, **31**, 742 (1988).
41. P. L'ECUYER, CACM, **33**, 85 (1990).
42. D. LE GALL, CACM, 34, 46 (1991).
43. V.O.K. LI, X. QIU, Proc. IEEE, **83**, 1210 (1995).
44. K. PAWLIKOWSKI, ACM Computing Surveys, **22**, 123 (1990).
45. C.H. SAUER, E. MACNAIR, Simulation of Computer Communication Systems (Prentice-Hall, Inc., Englewood Cliffs, NJ, 1983).
46. J.G. SHANTHIKUMAR, R.G. SARGENT, Oper. Res. **31**, 6, 1030 (1983).
47. H. TAKAGI, ACM Comp Surveys, **20**, 5 (1988).
48. S. TASAKA, Performance Analysis of Multiple Access Protocols (The MIT Press, 1986).
49. H.R. VAN AS, J.W. WONG, P. ZAFIROPULO, "Fairness, Priority and Predictability of the DQDB MAC Protocol Under Heavy Load", in Proc. 1990 Zurich Seminar, (Zurich 1990), 410-417.
50. J.W. WONG, J.P. SAUVE', J.A. FIELD, IEEE Trans. Commun., **30**, 346 (1982).

CHAPTER 7

AN IMPROVED MODEL OF HETEROGENEOUS ELEVATOR (SCAN) POLLING

Michael S. Borella and Biswanath Mukherjee

7.1. INTRODUCTION

An elevator (SCAN) polling system is one in which the server visits a group of N queues in the order $0, 1, 2, ..., N-2, N-1, N-2, N-3, ..., 2, 1, 0, 1, 2, ...$ and so on. While cyclic polling systems have been studied extensively[1,2], the analytical models of elevator polling have been limited in scope. This is largely due to the fact that these systems are very difficult to characterize in a compact manner.

This chapter introduces the first exact analysis of finite-buffer heterogeneous elevator polling. A Stochastic Activity Network (SAN) model is used to obtain numerical results. The research presented here is a generalization of Bunday, Sztrik and Tapsir[3], and is the first application of a SAN-based technique to solve an elevator polling system. The SAN model is general and heterogeneous, i.e., it can be easily modified to accommodate arbitrary buffer sizes, packet lengths, switching times, and arrival rates at each station. This is particularly useful in optimizing a system, such as this one, in which stations receive unequal amounts of bandwidth.

7.2. BACKGROUND AND RELATED WORK

7.2.1. Background

Early research on elevator polling grew out of machine room studies. Given a row of N machines (which periodically break) and an agent who patrols up and down the row (fixing machines when necessary), it is desirable to derive the efficiency of this polling technique. Later, elevator polling was studied as it applied to read/write head scheduling on computer disks, and to computer polling networks. Currently, it is a promising candidate for the analysis of multi-channel optical network protocols in which an etalon-based tunable receiver[4] scans a set of adjacent wavelengths. The IBM Rainbow prototype wavelength-division multiplexed network[5] employs such a protocol. The technique introduced in this chapter is being used to analyze the performance of that system.

7.2.2. Related Work

Author(s)	Buffer	Packet	Switching	Arrivals	Service	Hetero	Numerical Results
Swartz[6]	infinite	constant	general	general	gated	no	no
Coffman et. al.[7]	infinite	constant	constant	Poisson	exhaustive	no	no
Takagi et. al.[8]	infinite	constant	general	general	exhaustive	no	yes
Coffman et. al.[9]	one	constant	constant	Poisson	N/A	no	yes
Altman et al.[10]	infinite	general	general	Poisson	globally gated	yes	no
Bunday et al.[3]	one	geometric	constant	Poisson	N/A	yes	yes
Borella et. al.	finite	general	general	Poisson	any	yes	yes

Table 7.1: Previous studies of elevator polling.

Previous research in elevator polling is summarized in Table 7.1. Each study is categorized based on several characteristics. They are explained below:

- *Buffer*: Size (in packets) of the buffer at each node.

- *Packet*: Distribution of packet length.

- *Switching*: Distribution of switching time between stations.

- *Arrivals*: Arrival process at each station.

- *Service*: Service discipline used at each node (gated service is described in Takagi[2] and the globally gated discipline is discussed in Altman, Khamisy and Yechiali[10]).

- *Hetero*: Whether or not the model allows for stations to be heterogeneous (heterogeneous systems allow for different buffer sizes, packet length distributions, switching time distributions and arrival processes at each station).

- *Numerical Results*: Whether or not the study presented numerical results.

Note that it is intractable to produce numerical results unless exponential or constant distributions are used for the packet length and switching time. It is also usually intractable to produce numerical results when arrival processes other than Poisson are used. All of the papers in Table 7.1 that include numerical results make these simplifying assumptions, though several provide equations for the general cases as well.

Another difficulty in obtaining numerical results is that all elevator polling models proposed require very large state spaces (essentially, the entire state space must be enumerated, thus it grows exponentially with N). This limits the size of the system that can be studied numerically. For example, Takagi[1] reports results only for $N = 5$, and Bunday, Sztrik and Tapsir[3] provide results for $N = 8$ (they state that modern computing facilities can barely handle the case of $N = 10$).

7.3. OUR APPROACH

Developed in the early 1980's, Stochastic Activity Networks (SANs) allow high-level specifications of stochastic systems. They are based on stochastic extensions to Petri net modeling. SANs have been used to solve communication networks and fault tolerance modeling problems (see Sanders, Martinez, Alsafadi, and Nam[11] and Meyer and Wei[12] among others). A SAN model with exponentially distributed amounts of time spent in each state can be reduced to a Markov chain through an automated process, and subsequently solved for the steady state. There are software programs available that perform the reduction and solving, thus relieving the system designer from many hours of tedious work.

Figure 7.1: SAN primitives.

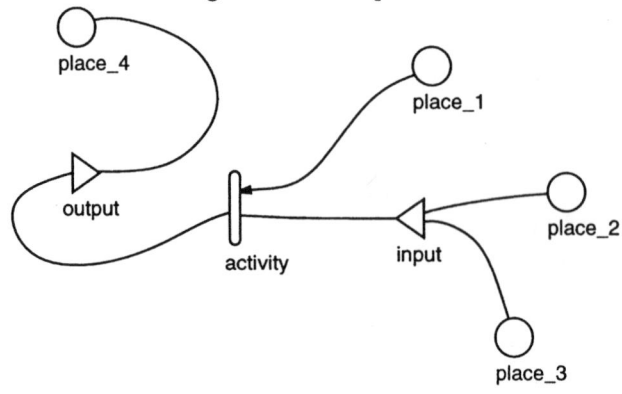

Figure 7.2: An example SAN.

SANs are composed of *places*, *activities*, and *gates* (see Figure 7.1). These primitives are connected by arcs, which describe the flow of control within the system. Places hold zero or more *tokens*. All tokens are identical – only the number of tokens in a place (also referred to as the *marking* of that place) is known. A marking of m tokens can either represent m items (such as the number of packets in a buffer) or a particular state (e.g., a marking of 1 token may represent a station that is idle, while a marking of 2 tokens may represent a station that is transmitting).

Places are connected to one another by activities. Activities can be *timed* or *instantaneous*. Activities are *fired* when all of the places connected via input arcs to that activity have at least one token in them, and all of the input gates of that activity have their predicates satisfied (see below). The length of the firing period of a timed activity is defined

Gate	Definition
input	Predicate $MARK(place\_2) == 0$ && $MARK(place\_3) == 2$
	Function $MARK(place\_2) = 1;$ $MARK(place\_3) --;$

Table 7.2: Input Gate Definitions for example SAN.

Gate	Definition
output	$MARK(place\_4) = 3;$

Table 7.3: Output Gate Definitions for example SAN.

by the distribution function of that activity (e.g., constant, exponential, normal, etc.). Instantaneous activities take a negligible amount of time to fire. Once an activity has completed firing, a token is removed from each place that is directly connected to the activity via input arcs, and a token is deposited in a place that is directly connected to the activity via an output arc.

All gates can only be defined between a place and an activity. *Input* gates control when an activity fires and can change the marking of the places connected to it. An input gate will not allow an activity to fire until the gate's *predicate* (a boolean function based on the marking of connected places) is satisfied. Input gates also have associated functions, which may change the markings of the places connected to the input gate. *Output* gates also have functions, but no predicates. Output functions can change the markings of places connected to the output gate. Upon the completion of a firing, all input and output functions are executed.

An example SAN is displayed in Figure 7.2 and its input and output gates are defined in Tables 7.2 and 7.3, respectively. The predicates and functions of gates can be specified with simple C language commands. In this SAN, activity *activity* will fire when there is at least one token in *place_1* and the predicate of *input* is satisfied, viz. *place_2* contains no tokens and *place_3* contains two tokens. After *activity* fires, the *place_2* is marked with one token, *place_3* is marked with one less token, and *place_4* is marked with 3 tokens.

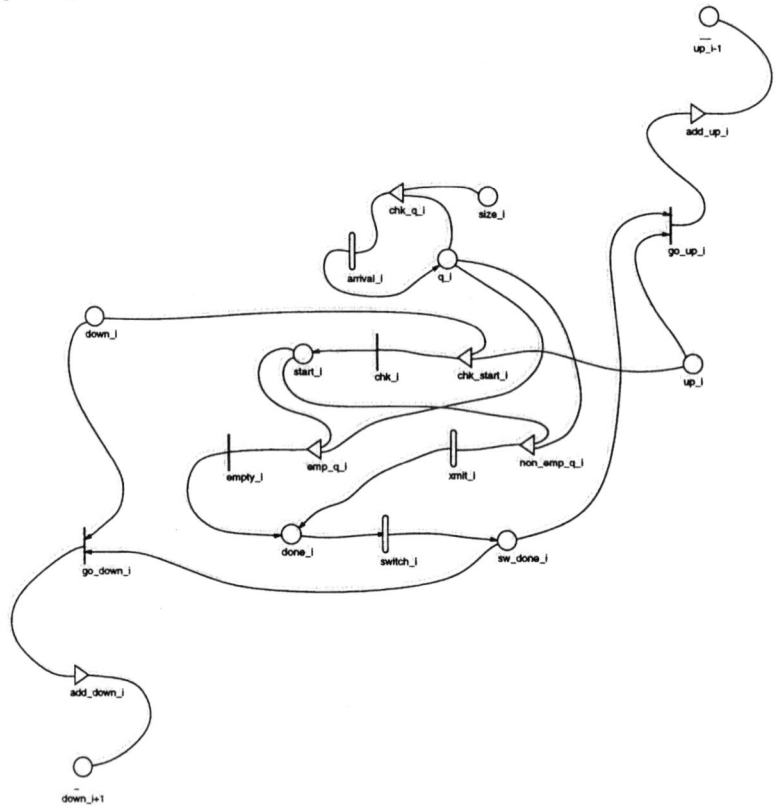

Figure 7.3: SAN for a single station of the elevator polling model.

The elevator polling model consists of N stations and the function of each is defined by a SAN. The path of the server (which represents the polling device that is moving from station to station) "connects" these stations to one another. Figure 7.3 and Tables 7.4 and 7.5 show how a partial SAN can specify a generic station i, $0 < i < N-1$.

Size_i is marked with the buffer size of station i. This marking never changes. When the number of tokens in station i's buffer (q_i) is less than the buffer size, the activity *arrival_i* will fire. After an amount of time determined by *arrival_i*'s distribution function, a packet will

AN IMPROVED MODEL OF HETEROGENEOUS ELEVATOR... 149

Gate	Definition
chk_q_i	Predicate $MARK(q\_i) < MARK(size\_i)$
	Function do nothing y
chk_start_i	Predicate $MARK(up\_i) == 2 \;\|\| \; MARK(down\_i) == 2$
	Function if $(MARK(up\_i) == 2) \; MARK(up\_i) = 1;$ if $(MARK(down\_i) == 2) \; MARK(down\_i) = 1;$
emp_q_i	Predicate $MARK(q\_i) == 0 \;\&\&\; MARK(start\_i) == 1$
	Function $MARK(start\_i) = 0;$
$non\_emp\_q\_i$	Predicate $MARK(q\_i) >= 1 \;\&\&\; MARK(start\_i) == 1$
	Function $MARK(start\_i) = 0;$ $MARK(q\_i) --;$

Table 7.4: Input Gate Definitions for station i's SAN.

be added to station i's buffer. The arrival of the server at station i is represented by a marking of two in either $down\_i$ or up_i (depending of which way the server is heading). When either of these two places has a marking of two, one of the tokens is immediately transferred to $start\_i$ by input gate chk_start_i. If the buffer is empty, viz. the marking of q_i is zero, the token in $start\_i$ is instantaneously transferred to $done\_i$. If the buffer is non-empty, activity $xmit\_i$ is fired (the server processes a packet), after which a token is removed from q_i and the token in $start\_i$ is transferred to $done\_i$.

Once a token is in $done\_i$, activity $switch\_i$ is activated. This activity represents the switching time of the server. When $switch\_i$ completes firing, a token is placed in sw_done_i. At this point, the system has to determine which direction the server has proceeded in – up or down. Instantaneous activity go_down_i will be fired when both $down\_i$ and sw_done_i have at least one token each. If the server was on its way down, one token was left in $down\_i$; thus, go_down_i will fire, and two

Gate	Definition
add_down_i	$MARK(up\_i-1) = 2;$
add_up_i	$MARK(down\_i+1) = 2;$

Table 7.5: Output Gate Definitions for station i's SAN.

tokens will be deposited into place *down_i+1*. Similarly, if the server was on its way up, *go_up_i* will fire and two tokens will be deposited into place *up_i-1*. Note that once the server leaves station i, all of station i's places will have a marking of zero, except for *q_i* and *size_i*.

The partial SAN described in this section can be linked with similar SANs to form stations 1 through $N-2$. Due to the fact that stations 0 and $N-1$ are at the end of the server's walk in each respective direction, a simpler representation of these end stations can be defined. Figure 7.4 displays the entire SAN model for an eight-station elevator polling network.

Numerical results are computed by solving the steady state probabilities of the system. Let $P_i(k)$ equal the fraction of time that there are i tokens in place k when the system is at steady state. Furthermore, let b_i equal the buffer size (in packets) of node i. Throughput (TP), blocking probability (BP) and mean delay (MD) can be found (per node) as follows:

$$TP_i = P_1(\text{start}\_i) \quad (0 \leq i \leq N-1)$$

$$BP_i = P_{b_i}(\text{q}\_i) \quad (0 \leq i \leq N-1)$$

$$MD_i = \frac{\sum_{j=0}^{b_i} jP_j(\text{q}\_i)}{TP_i} \quad (0 \leq i \leq N-1)$$

Mean delay will be normalized to the mean size of a packet.

7.4. RESULTS

The distribution functions of the arrival, service (transmission) and switching processes can be arbitrary. However, the model cannot be

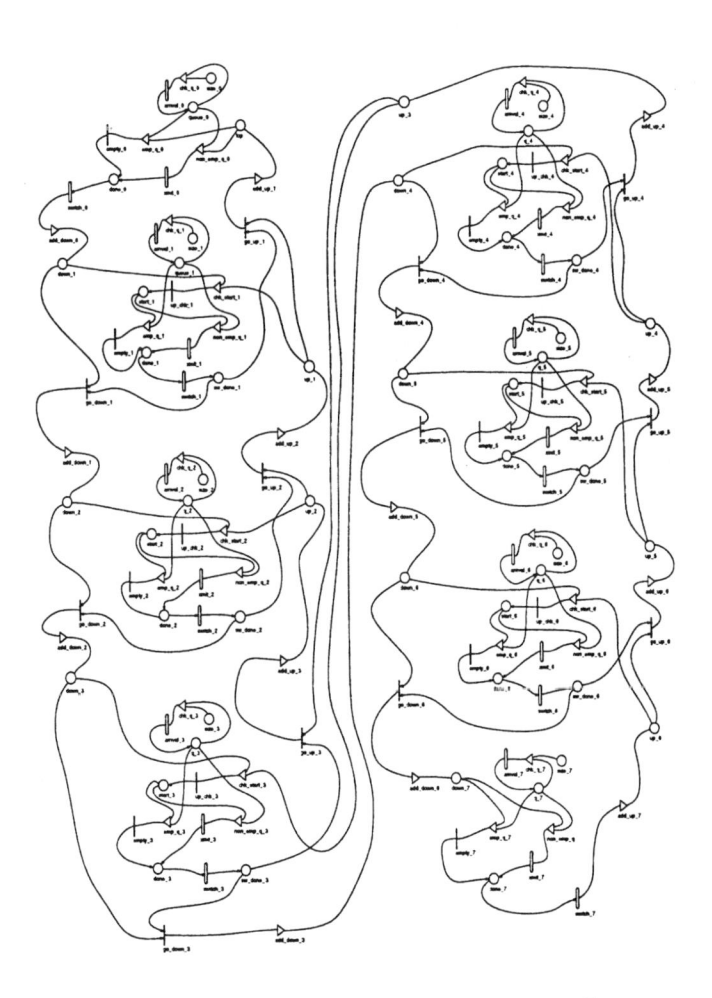

AN IMPROVED MODEL OF HETEROGENEOUS ELEVATOR...151

Figure 7.4: SAN for an eight-station elevator polling system.

analytically solved unless the arrival process is Poisson, and the service and switching processes are either constant or exponentially distributed. In the case that all three processes are exponentially distributed, the SAN can be reduced to a Markov chain and solved by conventional techniques. If the service and/or switching time is constant, the system can be iteratively solved for steady state through uniformization and successive overrelaxation[13].

This model was defined and solved using the *UltraSAN*[14,15] performability modeling tool. Numerical results were generated under the following assumptions:

- Messages arrive at station i according to a Poisson process with intensity λ_i, and are buffered in a FIFO queue.

- The length of packets arriving at station i is exponentially distributed with mean $1/\mu_i$.

- The buffer size at station i is b_i packets.

- Upon arrival at station i, the server will completely service exactly one packet before switching to the next station.

- The switching time between stations is a constant r.

Note that the model in Figure 7.4 can easily be modified to handle the cases of constant length packets and/or exponentially distributed switching times. Also, arbitrary service disciplines (such as limited, gated, globally gated, exhaustive, etc.) can be evaluated by modifying this model.

Analytical results were verified via independent simulation. While the *UltraSAN* package allows simulation of SANs, the simulation results in this section were generated by programs written by the authors.

Figures 7.5-7.7 show the throughput per node, blocking probability per node, and mean delay per node, respectively, for the parameters $N = 8$, $b_i = 1$ $(0 \leq i \leq 7)$, $\mu_i = 1$ $(0 \leq i \leq 7)$, and $r = 1$. Offered load $(\rho = \sum_{j=0}^{N-1} \lambda_j)$ is varied. The analytical and simulation results match extremely well. Note that the maximum overall utilization of this system is 50%, due to the fact that the mean service time is equal to the switching time. As expected, the closer to the center of the polling order that a station is, the more bandwidth it will receive; thus, its blocking probability and mean delay are reduced. The stations at the

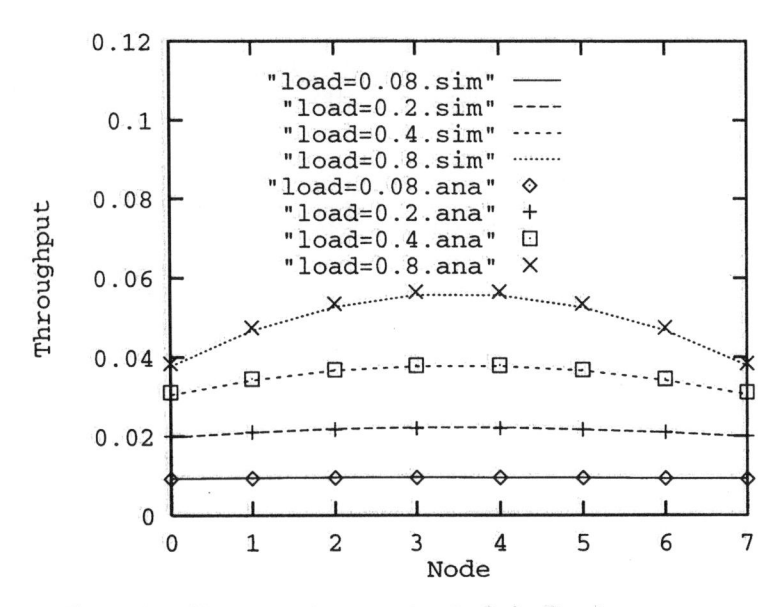

AN IMPROVED MODEL OF HETEROGENEOUS ELEVATOR...153

Figure 7.5: Throughput per node, single-buffer, homogeneous.

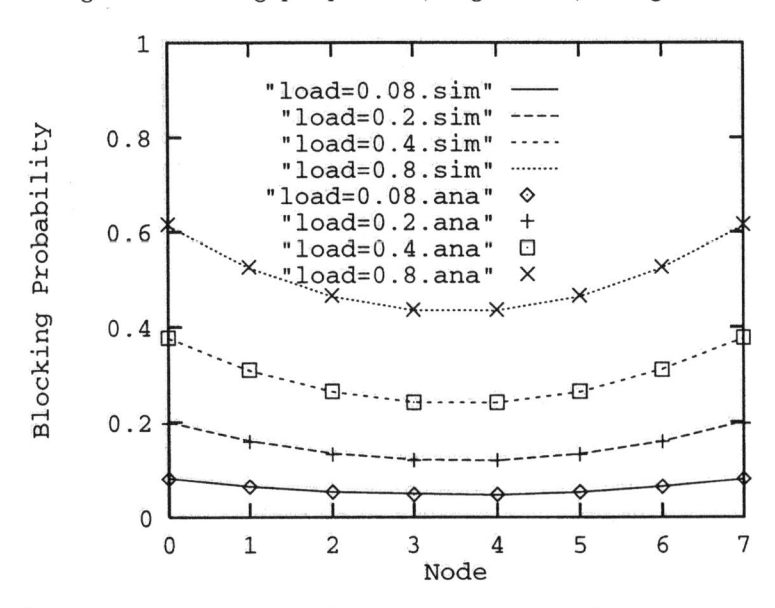

Figure 7.6: Blocking probability per node, single-buffer, homogeneous.

ends (stations 0 and 7 in this case) receive only one visit per complete cycle, whereas all other stations receive two visits. This is reflected in the numerical results.

In Figures 7.8-7.10, the four-node multiple-buffer case is considered for an offered load of 20% ($\lambda_i = 0.05$ ($0 \leq i \leq 3$), $\mu_i = 1$ ($0 \leq i \leq 3$), and $r = 1$). Buffer size is varied. Note that analytical results are only given for buffer sizes of one, two and four (see Section 2) due to the large state space that multiple-buffer models require. As the buffer size is increased, throughput is balanced, and the blocking probability decreases (it is negligible for buffer sizes greater than eight). Of course, mean delay is still slightly higher for the end nodes.

Figure 7.11 displays throughput per node for a heterogeneous system ($N = 4$, $b_0=b_3=3$, $b_1=b_2=2$, $\mu_i = 1$ ($0 \leq i \leq 3$), $r = 1$). By increasing the buffer sizes at the end nodes, the bandwidth allocated per node is more balanced for low to medium load. The ability to analyze heterogeneous systems opens the door to modeling polling under more realistic assumptions (e.g., client-server and/or nonuniform traffic patterns).

7.5. SUMMARY AND CONCLUSIONS

A limitation of this analytical approach (and every other analytical approach that has been suggested for this type of system) is that the model's state space grows exponentially with the number of stations and the buffer size at each station. The eight-station model requires the solution of a 5360 state model. Unfortunately, the eight-station multiple-buffer case (even for a buffer size of two) requires over 100,000 states, making it intractable to solve on mid-range workstations.

In the homogeneous case, it may be possible to use N SAN station specifications (like that of Figure 7.3) to approximate the performance of a $2N$ station system. Noting that the system characteristics will always be symmetrical (e.g., stations 0 to $N/2 - 1$ will show the same performance results as stations $N - 1$ to $N/2$, respectively), why even include stations 0 to $N/2 - 1$ in the model? The four station model was modified to estimate the eight station model by connecting place $down\_4$ to place up_4 with a switching time activity in between. This approximate model is reasonably accurate for systems with low offered

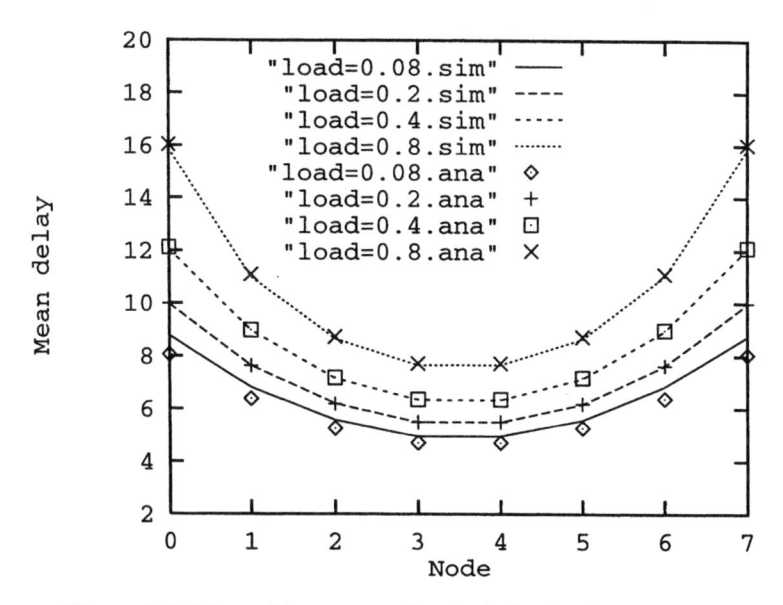

Figure 7.7: Mean delay per node, single-buffer, homogeneous.

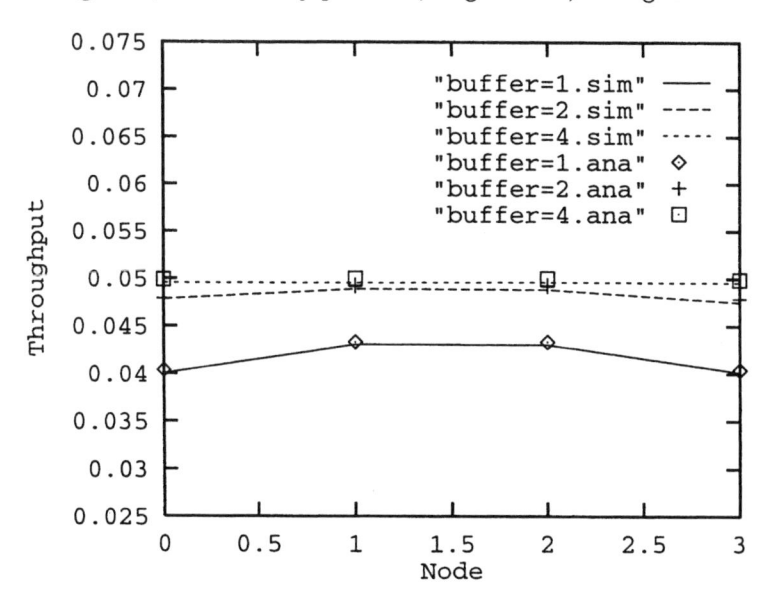

Figure 7.8: Throughput per node, load = 0.2, homogeneous.

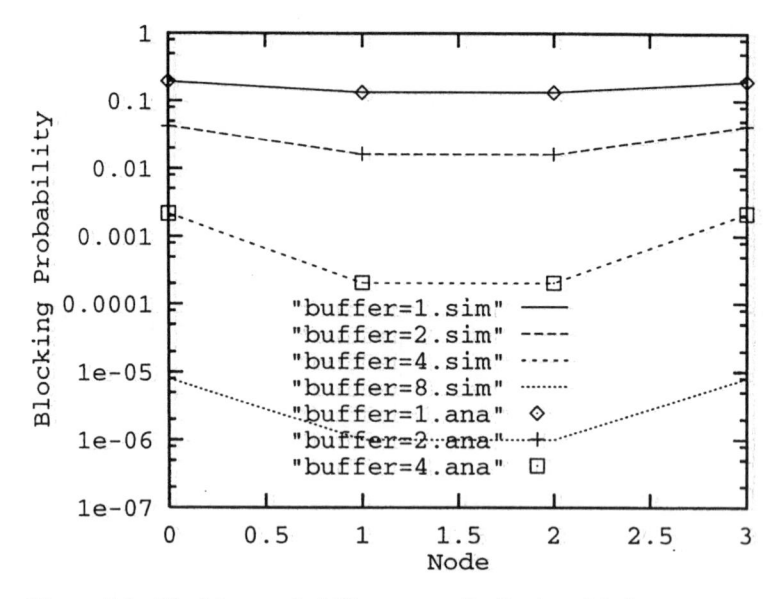

Figure 7.9: Blocking probability per node, load = 0.2, homogeneous.

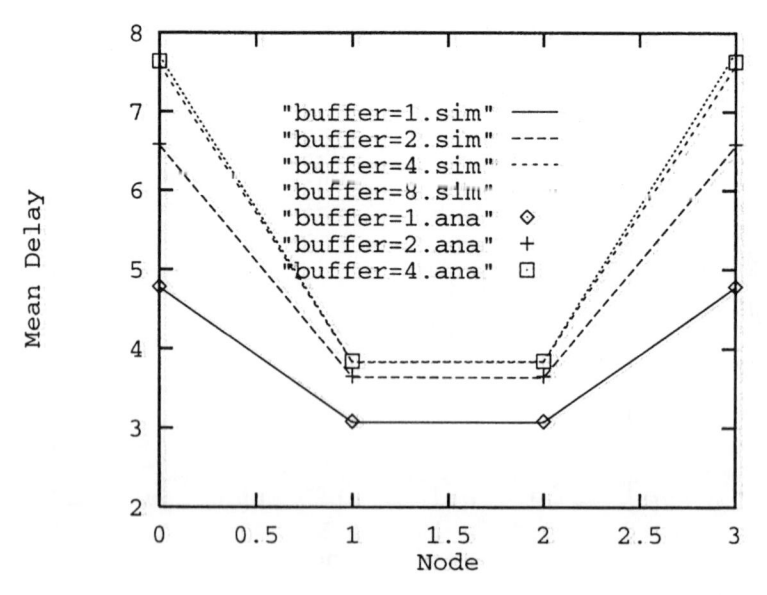

Figure 7.10: Mean delay per node, load = 0.2, homogeneous.

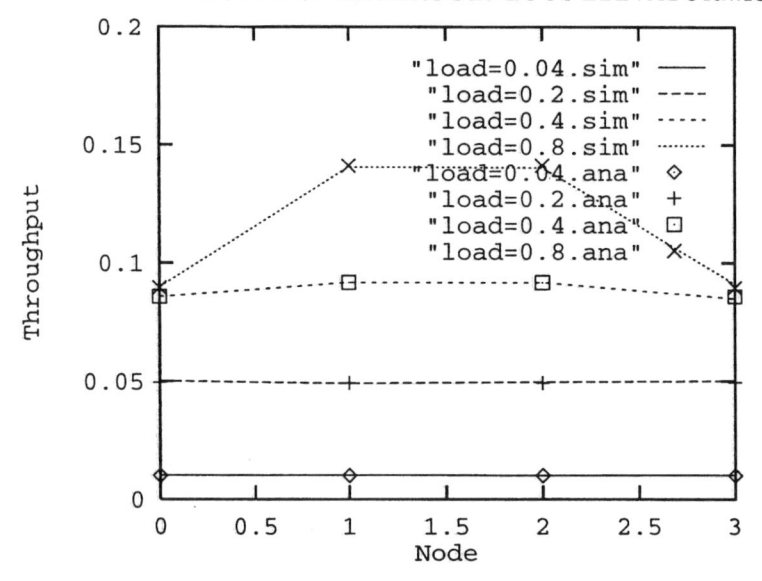

[Figure: Throughput vs Node plot with curves for load=0.04, 0.2, 0.4, 0.8 (sim and ana)]

Figure 7.11: Throughput per node, $b_0=b_3=3$, $b_1=b_2=2$, heterogeneous.

load[1] ($\rho < 25\%$). Work continues on improving this model, and accurately analyzing the performance of larger elevator systems remains an open problem.

The heterogeneous case allows an optimization problem to be defined. Given the traffic characteristics and buffer size of each station, how should the stations be placed so that mean delay or blocking probability is minimized or throughput is maximized? This also remains an open problem.

This chapter has presented a simple and concise method of achieving an exact analysis of a multiple-buffer heterogeneous elevator polling system through the use of Stochastic Activity Networks. This is the first study to solve this problem, and the solution has been verified via simulation. The scope of the model is limited to solving systems with less than about 10 nodes for the single-buffer case, and fewer nodes for the multiple-buffer case, however new techniques being developed may allow reasonable approximations of much larger systems. The SAN model described here is very general, and can be modified to produce exact results for any service discipline. We anticipate that this approach can

---

[1] Results for the approximation model are not given here due to space constraints.

be extended to analyze the performance of multiple-channel networks such as IBM's Rainbow optical network prototype, in which an etalon-based tunable filter is employed by each node to scan a set of adjacent wavelength channels.

REFERENCES

1. H. TAKAGI, *Analysis of Polling Systems*. (MIT Press, 1986).

2. H. TAKAGI, *Queueing Analysis of Polling Models: An Update*, in Stochastic Analysis of Computer and Communication Systems, (Elsevier Science Publishers, 1990), pp. 267-318.

3. B. D. BUNDAY, J. SZTRIK and R. B. TAPSIR, *A Heterogeneous Scan Service Polling Model with Single-Message Buffer*, in Performance of Distributed Systems and Integrated Communication Networks, (Elsevier Science Publishers, 1992), pp. 99-111.

4. H. KOBRINSKI and K.-W. CHEUNG, "Wavelength-Tunable Optical Filters: Applications and Technologies," IEEE Communications, **27**, (October 1989), pp. 53-63.

5. N. R. DONO, P. E. GREEN, K. LIU, R. RAMASWAMI, and F. F.-K. TONG, "A Wavelength-Division Multiple Access Network for Computer Communication," IEEE Journal on Selected Areas in Communications, **8**, (August 1990), pp. 983-993.

6. G. B. SWARTZ, *Analysis of a SCAN Policy in a Gated Loop System*. in Applied Probability – Computer Science: The Interface, Vol. 2, (Birkhauser, 1982).

7. E. G. COFFMAN and M. HOFRI, "On the Expected Performance of Scanning Disks," SIAM Journal of Computation, **11**, 1982.

8. H. TAKAGI and M. MURATA, *Queueing Analysis of Scan-Type TDM and Polling Systems*, in Computer Networks and Performance Evaluation, (Elsevier Science Publishers, 1986), pp. 199-211.

9. E. G. COFFMAN and E. N. GILBERT, "Polling and Greedy Servers on a Line," Queueing Systems, **2**, 2, (1987), pp. 115-145.

10. E. ALTMAN, A. KHAMISY and U. YECHIALI, "On Elevator Polling with Globally Gated Regime," Queueing Systems, 11, 1/2, (1992), pp. 85-90.

11. W. H. SANDERS, R. MARTINEZ, Y. ALSAFADI, and J. NAM, "Performance Evaluation of a Picture Archiving and Communication Network Using Stochastic Activity Networks," IEEE Transactions on Medical Imaging, 12, (March, 1993), pp. 19-29.

12. J. F. MEYER, and L. WEI, "Influence of Worload on Error Recovery in Random Access Memories," IEEE Transactions on Computers, 37, 4, 1988.

13. B. P. SHAH, "Analytic Solution of Stochastic Activity Networks with Exponential and Deterministic Activities," (Master's Thesis, University of Arizona, Department of Electrical and Computer Engineering, 1993).

14. J. COUVILLION, R. FRIERE, R. JOHNSON, W. D. OBAL II, M. A. QURESHI, M. RAI, W. H. SANDERS, and J. E. TVEDT, "Performability Modeling with UltraSAN," IEEE Software, 1991.

15. *UltraSAN Version 1.2.0 User's Manual*, (Performability Modeling Research Laboratory, University of Arizona, 1993).

CHAPTER 8

SIMULATION MODELING OF WEAK-CONSISTENCY PROTOCOLS

Richard A. Golding and Darrell D. E. Long

8.1. INTRODUCTION

Weak-consistency replication protocols can be used to build wide-area services that are scalable, fault-tolerant, and useful for mobile computer systems. We have developed the *timestamped anti-entropy* protocol. In it, pairs of replicas periodically exchange update messages; in this way updates eventually propagate to all replicas. We present results from a detailed simulation analysis of the fault tolerance and the consistency provided by this protocol. The protocol is extremely robust in the face of site and network failure, and it scales well to large numbers of replicas.

We have developed an architecture for building wide-area distributed services that emphasizes scalability and fault tolerance. This allows applications to respond gracefully to changes in demand and to site and network failure. It uses a single mechanism—weak-consistency group communication—to support both wide-area services and mobile computing systems.

The architecture uses data *replication* to meet availability demands and enable scalability. A group of servers or replicas cooperate to form a service. The replication is dynamic in that new servers can be added or removed to accommodate changes in demand. The system is asynchronous, and servers are as independent as possible; it never requires synchronous cooperation of large numbers of sites. This improves its ability to handle both communication and site failures. It uses large numbers of replicas, so replicas can be placed near clients and to spread query load over many sites.

We use *group communication* as the structuring principle for the architecture. A group consists of a set of *principals,* or group members, which communicate among themselves to coordinate a replicated service. The group members act as servers in a replicated service. They follow a *group communication protocol* when doing so, which provides a multicast service from one group member to the entire group. Clients can also multicast messages to the group, and the underlying protocol ensures that the communication reaches the appropriate group members.

Existing strong-consistency group communication protocols only work well with small groups. As the number of members increases, availability and response time decrease beyond the level acceptable for interactive applications, because they require the synchronous cooperation of a large number of principals. Instead we have turned our attention to *weak-consistency* protocols.

When weak-consistency group communication protocols are used, updates are first delivered to one site, then propagated asynchronously to others. The value a server returns to a client read request depends on whether that server has observed the update yet. Eventually, every server will observe the update. This allows updates to be sent using bulk transfer protocols, which provide the best efficiency on high-bandwidth high-latency networks. It also allows the group communication protocol to mask many failures, obviating the need for explicit recovery protocols in many cases. Several information systems, such as the Xerox Clearinghouse system,[4] have used similar techniques.

SIMULATION MODELING OF WEAK-CONSISTENCY PROTOCOLS 163

We have developed the *timestamped anti-entropy (TSAE)* weak-consistency protocol. It builds on our work on frameworks for implementing and analyzing group communication protocols.[8] This protocol provides *reliable, eventual* message delivery, ensures that message logs can be purged reliably, and supports mobile computer systems.

Analytical modeling of replication and group communication protocols is generally difficult, particularly when many principals are involved. We, like many other researchers, adopted simulation—both discrete event and Monte Carlo—as the appropriate method for evaluating our wide-area architecture.

In the remainder of this section, we will justify why we believe weak consistency protocols to be necessary for wide area systems, and discuss the measures we used to evaluate the TSAE protocol. Next we present the protocol. We then present the details of its analysis, concentrating on message reliability, delivery latency, message traffic, and inconsistency among principals.

8.1.1. The wide-area network environment

Existing consistent replication protocols are unsuited for wide-area applications because of the latency, failure, and scale of those networks. Latency affects the response time of the application, and can vary from a few milliseconds, for two hosts connected by an Ethernet, to several hundred milliseconds for hosts on different continents communicating through the Internet. Packet loss rates often reach 40%, and can go higher.[6] The Internet has many points that, on failure, partition the network, and at any given time it is usually partitioned into several non-communicating networks. Further, in 1992 the Internet included more than a million hosts,[12] with a potential user base several times larger. This has led to query loads on some services exceeding the capacity of a single workstation and the network links that support it.[5] Any group communication protocol that requires interactive communication with many principals will not work in this environment.

The introduction of mobile computers exacerbates this problem further. These systems spend most of their time disconnected from other systems, or perhaps connected by an expensive or low-bandwidth radio link.

Services that are to support such systems must allow clients that are disconnected to continue to operate, without communicating with outside servers. This can be accomplished by placing a group member on the mobile system and allowing the copy to diverge from the "correct" value. Weak consistency protocols ensure that this divergence can be reconciled when the mobile system is reconnected to the network. The timestamped anti-entropy protocol allows mobile systems with limited bandwidth to measure how far they have diverged by exchanging a small summary of the state of the group member with another member.

Despite these restrictions, users expect a service to behave as if it were provided on a local system. The response time of a wide-area application should not be much longer than that of a local one. Further, users expect to use the service as long as their local systems are functioning.

We assume that principals have access to pseudo-stable storage such as magnetic disk that will not be affected by a system crash. Principals, or the hosts on which they run, have loosely synchronized clocks. Principals and hosts fail by crashing; that is, when they fail they do not send invalid messages to other principals and they do not corrupt stable storage. Hosts can temporarily fail and then recover. Principals have two failure modes: temporary recoverable failures, and permanent removal from service. The network is sufficiently reliable that any two principals can eventually exchange messages, but it need never be free of partitions. *Semi-partitions* are possible, where only a low-bandwidth connection is available.

We believe that weak-consistency protocols provide a good solution for building a replicated service on a wide-area network. Clients can communicate with the replica nearest them, reducing message traffic on the network and spreading query load over many servers. The TSAE protocol, in particular, scales well to large numbers of replicas. The protocols make many faults, such as network partitions and system crashes, invisible to the replication service by allowing copies to diverge, then reconciling them when the fault is repaired. This is a natural way of handling mobile, disconnected computer systems as well.

8.1.2. Measures

How well can weak-consistency protocols be expected to work? How do they compare to other approaches? We used several different performance measures to answer these questions.

Two of the measures concern individual messages. *Message reliability* measures how often the protocol will fail to deliver a message because principals are removed from service without notice, while *message latency* measures how long the protocol takes to deliver a message.

The other two measures aggregate the effects of many messages. *Message traffic* indicates how many messages applications using weak-consistency protocols will generate, and indirectly how they interfere with other network activity. *Consistency* measures how up-to-date each principal can be expected to be.

8.2. PROTOCOL DESCRIPTION

Replicated data can be implemented as a group of replica principals that communicate through a *group communication protocol*. The group communication protocol generally provides a *multicast* service that sends a message from one principal to all other principals in the group.

The protocol determines the *consistency* of each principal by controlling the order messages are sent among them. Consistency is often defined in terms of the value that each principal holds. However, that value is determined by the messages that the principal has received, so we define consistency in terms of messages. *Weak* consistency protocols guarantee that messages are delivered to all members but do not guarantee when.

8.2.1. Kinds of consistency

The consistency provided by a group communication protocol can be classified by the guarantees it provides. This includes guarantees on *message delivery, time of delivery,* and *delivery ordering*. In general, strong guarantees require multiphase synchronous protocols while weaker guarantees allow efficient asynchronous protocols.

Messages can either be delivered *reliably* or not. We have identified four levels of reliability: *atomic*, where the message is either delivered to every member, or to none; *reliable*, where the message is delivered to every functioning principal, or to none; *quorum*, where the message is delivered to at least some fraction of the membership; and *best-effort*, where delivery is attempted but not guaranteed. These levels differ in the way they handle failed principals and unreliable communication media.

Likewise, the communication protocol can deliver messages *synchronously*, within a *bounded* time, or *eventually* in a finite but unbounded time. This guarantee is orthogonal to reliability.

Messages will be delivered to principals in some order, perhaps different from the order in which they are received. There are several possible orders, including total, causal,[11, 10] and bound inconsistency.[13, 3] Weaker orderings include a *per-principal* or *FIFO channel* ordering, where the messages from any particular principal are delivered in order, but the streams of messages from different principals may be interleaved arbitrarily.

Weak consistency protocols provide reliable delivery, and can provide one of several delivery orderings, but only guarantee eventual message delivery. In particular, all principals are guaranteed to agree in finite but unbounded time if no further messages are sent.

These protocols are generally implemented using background message propagation. When a client wishes to update the value a group maintains, it sends a message to one group member. This member then forwards it to another member. Later both members forward it to other members, and so on until it has spread throughout the group.

Anti-entropy protocols use this technique, and guarantee reliable delivery. Principals using this protocol periodically exchange messages with other principals, potentially continuing this process forever. Other weak consistency protocols, such as rumor mongery,[4] do not guarantee reliable delivery, because principals use probabilistic rules to determine when to stop propagating a message.

8.2.2. Timestamped anti-entropy

We have developed a new group communication protocol called *timestamped anti-entropy*.[7] It provides reliable, eventual delivery, so that messages are delivered to every principal in the group even if some have temporarily failed or are disconnected from the network. We have also developed a related group membership mechanism that provides for adding and removing members from the group.[9] The trade-offs are that the protocol may have to delay message delivery until a fault is repaired, that members must maintain logs on disk that are not compromised by failure and recovery, and that timestamp information must be appended to every message.

We have developed the TSAE protocol, a new group communication protocol that provides reliable, eventual delivery.[7] Like other weak-consistency protocols, update messages originate at a single principal and are propagated in the background to others. Unlike other protocols, TSAE supports several message delivery orders, mobile computer systems, and provably correct purging of message logs. We have also developed a related group membership mechanism that handles adding and removing members from the group.[9]

When a principal wishes to send a message, presumably as the result of a client performing an update operation, the message is marked with a timestamp and the identity of the sending principal, and written to a message log. This log is maintained on stable storage, so that it survives temporary crashes. It is organized as a set of sub-logs, one for each principal in the group.

Each principal maintains two *timestamp vectors* on stable storage in addition to the message log. Each vector is indexed by principal identifiers. The summary vector maintains a summary of the messages in the log, and when it holds a timestamp value for a particular principal, all messages sent by that principal with lesser or equal timestamp have been received. The entry in the summary vector for principal P is always larger than the greatest timestamp recorded in the message sub-log for P.

The acknowledgment vector summarizes the messages that have been received by other principals. Since clocks are loosely synchronized, a principal can acknowledge that it has received all messages sent by any principal up to a particular time.*

From time to time, a principal will select another principal as its partner, and the two will exchange messages in an *anti-entropy session*. The session begins with the principals exchanging their summary vectors. Using this information, each principal can determine the messages in its log that its partner has not yet received. These messages are sent using a reliable stream communication protocol. The session ends with the principals exchanging their acknowledgment vectors.

When a session is complete, both principals have received exactly the same set of messages. Moreover, they have received a continuous sequence of messages from each principal. To see that this is so, consider the first time one principal, call it A, contacts another principal, B. Each principal will only have messages it has sent in its log, and its summary vector will contain zero timestamps except for its own entry. A and B will exchange messages, each receiving a continuous sequence of messages from the other. Some time later, A will contact B again and receive more messages sent by B. At the beginning of the session, the summary timestamp that A records for B will be earlier than the summary timestamp that B records for itself. B will forward all the messages that A has not yet received, and A will once again have a continuous sequence of messages from B in its log.

The acknowledgment vector is used to keep the log from growing without bound. At the end of an anti-entropy session, principal A finds the minimum timestamp, min, in its summary vector. A has received every message from any sender that has a lesser or equal timestamp. A implicitly acknowledges all these messages by setting its entry in its acknowledgment vector, acknowledgment[A], to min. A log entry can be purged when every

---

*We have also developed a similar protocol that requires $O(n^2)$ state per principal rather than $O(n)$, but allows unsynchronized clocks. This alternate protocol was discovered independently by Agrawal and Malpani.[1]

other process has observed it, which is true when the minimum timestamp in the acknowledgment vector is greater than the timestamp on the log entry.

There are several variations on the basic TSAE protocol. The timestamp vectors and message timestamps can be used to order messages before they are delivered from the log. Anti-entropy sessions can be augmented by a best-effort multicast to speed propagation, and principals can use various policies to select partners.

Best-effort multicast combined with anti-entropy will spread information rapidly. When a principal originates a message, it can multicast it to other principals. Some of them will not receive the multicast, either because the network dropped the message or because the principal was temporarily unavailable. These members will receive the message later when they conduct an anti-entropy session with another member that has received the message. This speeds dissemination when message delivery order is not important.

8.2.3. Partner selection

There are several possible policies for selecting a partner for an anti-entropy session. Table 1 lists eight of them. The TSAE protocol requires only that every neighbor eventually be contacted to ensure that messages are delivered reliably, and weaker constraints can work for some network topologies.

The policies can be divided into three classes: random, deterministic, and topological. Random policies assign a probability to each neighbor, then randomly select a partner for each session. The deterministic policies use a fixed rule to determine the neighbor to select as partner, possibly using some extra state such as a sequence counter. Topological policies organize the principals into some fixed graph structure such as a ring or a mesh, and propagate messages along edges in the graph.

Table 8.1. Partner selection policies

Random policies:	
Uniform	Every neighbor principal has an equal probability of being randomly selected.
Distance-biased	Nearby neighbors have a greater probability than more distant neighbors of being randomly selected. Should reduce traffic on long-distance backbone links.
Oldest-biased	The probability of selecting a neighbor is proportional to the age of its entry in the summary vector.
Deterministic policies:	
Oldest-first	Always selects the neighbor n with the oldest value in the summary vector.
Latin squares	Builds a deterministic schedule guaranteed to propagate messages in $\theta(\log n)$ rounds.[2]
Topological policies:	
Ring	Organizes the principals into a ring.
Binary tree	Principals are organized into a binary tree, and propagate messages randomly along the arcs in the tree.
Mesh	Organizes the principals into a two-dimensional rectangular mesh.

8.3. MESSAGE RELIABILITY

The timestamped anti-entropy message delivery protocol provides reliable eventual delivery. However, reliable delivery does not guarantee that a message is delivered when its sender fails. For the TSAE protocol to fail to deliver a message, every principal that has received a copy—which

SIMULATION MODELING OF WEAK-CONSISTENCY PROTOCOLS

includes the sender—must fail. This section examines how often this happens.

Delivery reliability can be measured by the probability that a message will be delivered to every principal before all recipients can fail. The probability is affected by the rate at which anti-entropy sessions propagate messages, and the rate at which principals fail.

In practice, the only way a principal can fail is in a sudden, catastrophic removal from service—a fire or hardware failure, for example. This sort of failure is extremely rare for systems on the Internet. The analysis in this section, however, explores a wide range of failure rates.

8.3.1. Monte Carlo simulation

Message loss can be modeled using a state transition system like that shown in Figure 1. Each state is labeled with a pair (m,f), where m is the number of functioning principals that have observed a message, and f is the total number of functioning principals. The system starts in state $(1,n)$, with one principal having observed a message out of n possible (5 in the example). The system can then either propagate the information using anti-entropy, in which case the system moves to state $(2,n)$, or a principal can be removed from service and the system moves into state $(0,n-1)$. The message has been lost when the system reaches a state $(0,x)$, and it is delivered when it reaches (x,x).

Anti-entropy and principal failure are treated as Poisson processes with rate λ_a and λ_f, because Poisson processes are easy to model and analyze. Real systems often follow more complex distributions, but this Poisson process modeling are known to be robust even if the underlying distribution is not exponential.

The rate of *useful* anti-entropy sessions, where a principal that has received the message contacts one that has not, is a function of m, f, and the partner selection policy. In particular, f principals will be initiating anti-entropy sessions. If principals choose their partners randomly, each

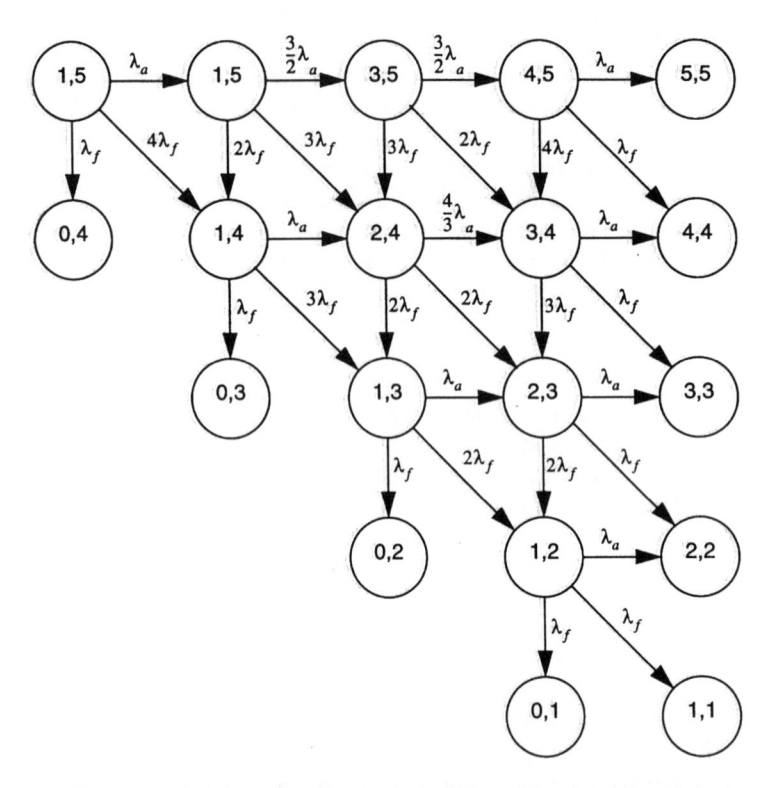

FIGURE 8.1. Model of message receipt and failure for five principals. This model only includes permanent failure; transient failure and recovery would add an additional dimension of states.

principal that has observed the update has a chance $(f - m)/(f - 1)$ of contacting a principal that has not yet observed the update. Since anti-entropy is a Poisson process, the rate of useful anti-entropy sessions is

$$m\frac{f-m}{f-1}\lambda_a.$$

Since removal from service is a permanent event, the state transition graph is acyclic, with $\theta(n^2)$ states in the number of principals. The probability p_i of reaching each state i can be computed using a sequential walk of the states. The probability density functions $p_i(t)$ of the time at which the system enters each state can be derived analytically or numerically. The analytic solution for $p_i(t)$ can be found by convolving the entry-time distribution $p_j(t)$ for each predecessor state j with the probability density of the time required for the transition from j to i. Alternately, the system can be solved numerically using a simple Monte Carlo evaluation.

8.3.2. Results

Figure 2 shows the probability of removal from service interfering with message delivery for different numbers of principals. The probability is a function of $\rho = \lambda_a/\lambda_f$, the ratio of the anti-entropy rate to the permanent site failure rate. The two graph is plotted using a linear vertical scale, which emphasizes the behavior for small values of ρ. The probability asymptotically approaches zero as ρ increases.

Internet sites generally are removed from service after several years of service, and then usually with enough notice to run a leave protocol. The anti-entropy rate is therefore likely to be many thousands of times higher than the permanent failure rate. As a result, there will be almost no messages lost because of removal from service.

8.3.3. Volatile storage

Some implementations may buffer messages in volatile storage before copying them to the stable message log. Write-back caches are used by many operating systems. These implementations will lose the information in volatile storage when a principal temporarily fails and recovers.

Volatile storage complicates the state transition model. States must be labeled with four values: the number of functioning principals that have not observed a message, the number that have written it to volatile store, the

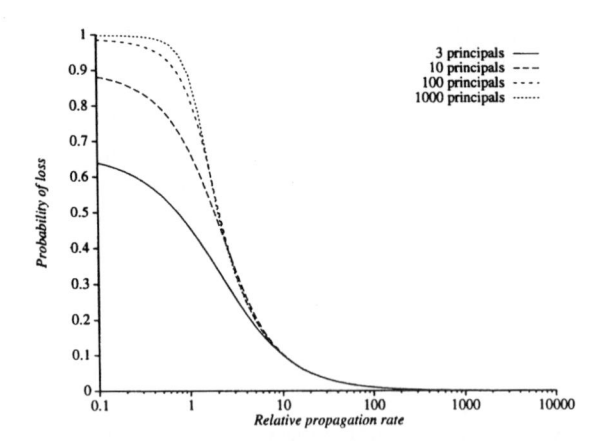

FIGURE 8.2. Probability of failing to deliver a message to all sites (linear vertical scale). The relative propagation rate ρ is the ratio of the anti-entropy rate λ_a to the permanent site failure rate λ_f. The linear scale emphasizes the effect of the number of principals for small values of ρ.

number that have written it to disk, and the number that have temporarily failed. The state transitions are complex and the solution is impractical for realistic numbers of principals.

However, the effect of volatile storage can be bounded by considering the probability that a failure will occur while there are messages that have not been made stable. Assume that temporary failure is a Poisson process with rate λ_t and that volatile data is flushed to stable storage every s time units. The probability that a failure occurs before writeback is

$$p = \frac{-2e^{-s\lambda_t} + s^2\lambda_t^2 - 2s\lambda_t + 2}{2s\lambda_t^2}$$

For a typical value of $s=30$ seconds and $1/\lambda_t = 15$ days, p is so close to zero as to be negligible.

8.4. MESSAGE LATENCY

The group communication protocol provides latency guarantees as well as reliability. The TSAE protocol only guarantees eventual delivery, but in practice messages propagate to every principal rapidly.

If information is propagated quickly, clients using different principals will not often observe different information, and loss of an update from site failure will be unlikely.

8.4.1. Simulation method

We constructed a discrete event simulation model of the TSAE protocol to measure propagation latency. The latency simulator measured the time required for an update message, entered at time zero to propagate to all available principals. The time required to send a message from one principal to another was assumed to be negligible compared to the time between anti-entropy sessions. The simulator could be parameterized to use different partner selection policies and numbers of sites. The simulator was run until either the 95% confidence intervals were less than 5%, or 10 000 updates had been processed. In practice 95% confidence intervals were generally between 1 and 2%.

The simulation modeled only the TSAE protocol, and did not consider the effect of combining TSAE with a best-effort multicast. Therefore the results in this section represent worst-cast behavior that would be improved if a multicast were added.

8.4.2. Results

The partner selection policy also affects the speed of message propagation. Figure 2 shows the mean time required to propagate a message to every principal for several policies as the number of sites increases. The uniform, latin squares, and distance-biased policies give essentially identical performance. Age-biased appears to provide slightly better performance, which would appear to contradict the claim by Alon et al. that the latin squares policy is fastest.[2] We believe the difference arises from a slight difference in implementation: Alon's implementation requires that every

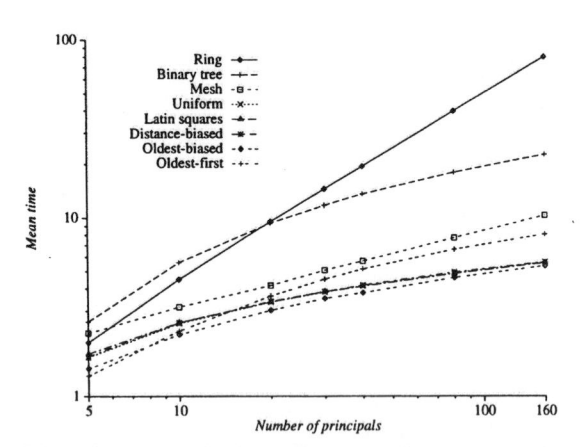

FIGURE 8.3. Effect of partner selection policy on scaling of propagation time.

principal propagate messages in well-defined rounds, while this simulation allows propagation to occur at random intervals. This may mitigate some of the benefit derived from Alon's latin squares policy. The policies that simulate a fixed topology—ring, mesh, or binary tree—have the worst performance and scaling.

These results indicate that simple random policies, such as uniform selection or age biasing, perform quite well.

8.5. TRAFFIC

A group communication system must not overload the network on which it operates. This is particularly important if the group is to scale to a large size. The *traffic* induced by a system, as measured by the number of network packets that are sent and by the distance the packets traverse, is the primary measure of how the system will affect the network.

Researchers at Xerox PARC found that the original version of the Clearinghouse system overloaded their internetwork.[4] The original implementation combined anti-entropy sessions with a best-effort multicast, and used a uniform partner selection policy during anti-entropy. A revised implementation reduced the network load using a distance-biased partner selection policy.

Network traffic is measured by the number of packets sent, and by how many network links they must traverse. The number of packets is determined by the number of messages each principal sends, their size, and how often principals perform anti-entropy. In the absence of principal or network failures, the TSAE protocol ensures that every message is sent exactly once to each principal. The number of links each packet must cross is determined by the topology of the network and by the partner selection policy that principals use when performing anti-entropy.

For this performance evaluation, we introduced two partner selection policies in addition to those discussed in Section 2. The cost-biased policies preferentially select low cost partners. This is different from distance-biased selection when network links have different costs. The cost-biased policy selects a principal with probability proportional to the inverse of the difference between its cost and the highest cost in the group. The cost-squared-biased selects with probability proportional to the difference squared, increasing the probability that low-cost partners will be selected. The advantage of these policies is that they are not based on an arbitrary topology, but rather upon observable performance measures. This makes them appropriate for use in a wide-area internetwork where topological information is likely to be unavailable.

8.5.1. Simulation method

The system was modeled in a discrete-event simulation, written using a locally-developed library of C++ classes. Principals initiated anti-entropy sessions according to a Poisson process. Each run of the simulation performed 1000 anti-entropy sessions, collecting link traffic and cost statistics. At least 100 runs were performed, and they were repeated either until 1000 runs completed or until a batch-means analysis indicated that the 95% confidence interval width for each measure was less than 1% of the measured value. A complete set of runs required between half an hour and six hours on a DECstation 5000/200, depending on the partner selection policy.

The simulator programs modeled a particular physical network topology. Principals were grouped into 5-cliques, and the cliques were connected by gateways to a backbone ring. This topology was selected not because it specifically modeled part of the Internet, but because it is representative of topologies that can concentrate traffic onto a few links. It is similar to the structure of the Internet today: regions of high connectivity, with regions weakly connected through a backbone network. Communication within a region is fast, while communication between regions can be expensive. Demers et al. noted that this kind of topology was a problem for the original Clearinghouse protocols.[4]

Each simulator run was parameterized by the partner selection policy. The binary tree and mesh partner selection policies were not tested because they were patently inappropriate on ring-like topologies. The number of principals was fixed at 30 (six 5-cliques) but the cost of the backbone links relative to the intra-clique cost could be specified. This cost ranged from 1 through 160. The upper bound was chosen arbitrarily; it represents more than an order of magnitude difference from the lowest cost.

8.5.2. Results

Figure 2 shows the average cost of the links traversed by an anti-entropy session. As on a simple ring, the uniform, oldest-biased, oldest-first, and latin squares policies all performed about the same, while distance-biasing improves performance somewhat. The ring policy is somewhat better than all of these, though it scales in about the same way. The cost-biasing protocols produce somewhat more traffic when the backbone cost is low, but scale better as the backbone cost increases.

The cost-biasing policies decrease the traffic per link as the cost of the backbone increases. Figure 2 shows that the cost decrease is achieved by concentrating communication within a clique. When backbone links cost 80 times as much as a clique link, the cost-biased policy only allows between 1 and 2% of all anti-entropy sessions to cross between cliques. Fewer than 0.3% of all sessions cross backbone links when cost-squared-biasing is used. Clearly cost-biasing ensures that communication is predominantly local.

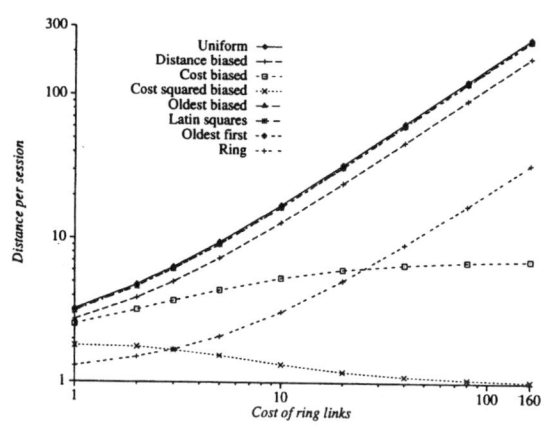

SIMULATION MODELING OF WEAK-CONSISTENCY PROTOCOLS 179

FIGURE 8.4. Effect of partner selection policy on the average number of network links used in an anti-entropy session. Measured on a ring of six 5-cliques. The cost of the backbone ring links varied from 1 to 160.

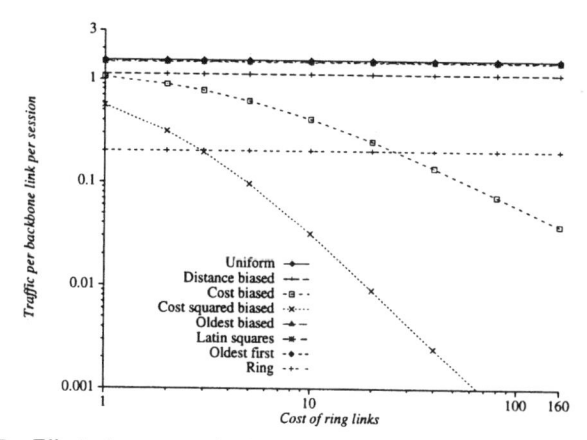

FIGURE 8.5. Effect of partner selection policy on the mean traffic per backbone ring link. The cost of the backbone links varied from 1 to 160. Only the cost-biased policies adapt to the cost of backbone links.

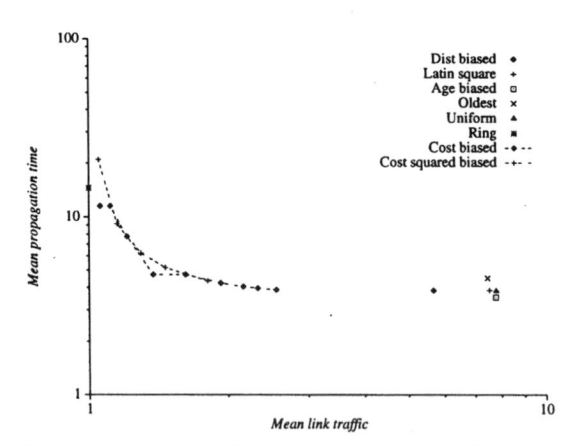

FIGURE 8.6. Scatter plot of the relationship between link traffic and propagation delay. Measured for 30 principals. Time is measured in multiples of mean time between anti-entropy sessions.

While one might want to reduce network traffic as much as possible, a reduction in traffic generally requires an increase in the time required to propagate a message. Figure 2 shows the relationship between the two. The time required to propagate a message is plotted as a function of the link traffic for each of several partner selection protocols. Every policy except cost-biased and cost-squared-biased is represented by a single point. Multiple points are reported for the cost-biased policies, showing how they respond to different backbone link costs. All of the policies excepting the ring policy appear to fit on a smooth curve.

8.6. CONSISTENCY

Weak consistency protocols allow principals to contain out-of-date information. There are two related measures of this effect—one concerning the propagation latency of a single message, the other concerning the consistency of group state. The likelihood that a principal is out-of-date with respect to other principals, and the difference between them, aggregates the effects of several messages.

8.6.1. Simulation method

A discrete event simulation modeled the TSAE protocol to measure information age. The simulator used five events: one each to start and stop the simulation, one to send a message, one to perform anti-entropy, and one to sample the state of a principal. The simulation was first allowed to run for 1 000 time units so it would reach steady state, then measurements began. The simulation ended at 50 000 time units. Read, write, and anti-entropy events were modeled as Poisson processes with parameterizable rates. These rates were measured per principal. The simulator included different partner selection protocols and an optional unreliable multicast on writes.

The simulator maintained two data structures for each principal: the anti-entropy summary vector and a message number. It also maintained a global message counter. When a message was sent, the global counter was incremented and the sender's message number was assigned that value. If an unreliable multicast was being used, the message number would be copied to other principals if a the datagram was received. Anti-entropy events propagated message numbers between principals, as well as updating the principals' summary vectors.

Sampling events were used to collect measures of the expected age of data and the probability of finding old data. A principal was selected at random, and the message number for that principal was compared to the global counter. The difference showed how many messages the principal had yet to receive.

Our intent is to measure the consistency of *entire* databases. We measure the age of the state of a principal by the number of messages it has yet to receive. We expect most wide-area services to provide large databases, and so the messages will contain updates to many different database entries. The expected age, therefore, must be divided by number of data entries to arrive at the maximum probability that any particular entry is out-of-date.

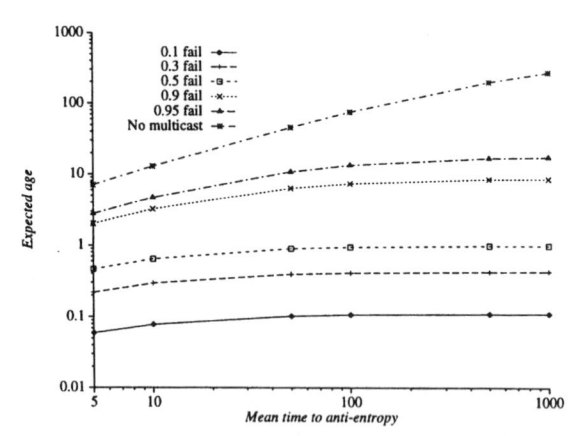

FIGURE 8.7. Expected data age as anti-entropy rate varies, for 500 principals. Mean time-to-write 1 000; uniform partner selection. Anti-entropy was combined with a best-effort multicast. The different curve show the effect of the probability of multicast message failure.

The age of a principal's state depends on the ratio of the anti-entropy rate to the update rate for the state. The graphs in this section were generated using a mean time-to-update of 1 000 time units; the maximum anti-entropy rate investigated was only 200 times greater, giving a mean time-to-anti-entropy of five. This implies that all the results presented here are more pessimistic than would actually be observed.

8.6.2. Results

Our first analysis investigated the age of the information held by principals, and the degree to which it is affected by varying the rate of anti-entropy. Figure 2 shows the expected age of a member's state. Clearly, adding an unreliable multicast on write significantly improves consistency. The message success probability is the most important influence on information age in large groups of principals. For small numbers of principals, increasing the anti-entropy rate dramatically improves both the probability of getting up-to-date information and the expected age.

Figure 2 shows how consistency depends on the number of principals. For these simulations the anti-entropy rate was fixed at 100 times that of writes. This value might be typical for a system where entries are corrected soon after being written. Later updates will be less frequent and the ratio will increase, improving the consistency. Once again an unreliable multicast provides considerable improvement.

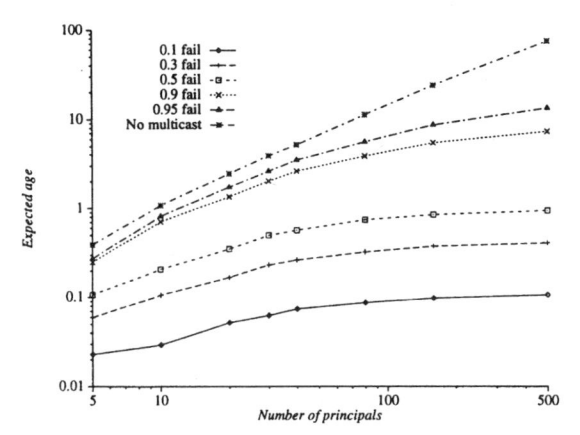

FIGURE 8.8. Expected data age as the number of principals varies, with anti-entropy occurring 100 times as often as writes. Uses uniform partner selection. Also shows the effect of varying message failure rates in a best-effort multicast.

We also investigated the effect of partner selection policy on information age. The results ranked partner selection policies in exactly the same order as they were ranked for propagation latency (Figure 2). The topological policies (ring, binary tree, and mesh) propagate more slowly, and give a greater expected age, than other policies. The other policies are nearly equal, though oldest-first has a slight advantage.

8.7. CONCLUSIONS

We have evaluated the performance of the *timestamped anti-entropy* (TSAE) protocol, including network traffic, message latency, fault tolerance, and consistency. Simulation has proven a useful mechanism for investigating these goals. It would have been difficult to build and measure a system on the Internet with hundreds of replicas, particularly if we wanted to evaluate the protocols under specific failure conditions.

The analysis shows that the TSAE message delivery protocol scales well to large groups of principals. The time required to propagate an update from one principal to all others increases as the log of the size of the group, and the partner selection policy can be chosen to control network traffic as the membership grows.

Fault tolerance is the ability of a system to provide correct service even when some parts fail. The TSAE protocol provides excellent fault tolerance by delaying communication until a principal is available.

The negative aspect of weak consistency protocols is that principals will have out-of-date information. This investigation found that an unreliable multicast can mitigate most of this problem, and that at reasonable propagation rates principals are rarely more than a few updates behind.

The timestamped anti-entropy protocol meets our goals of scalability, reliability, and fault tolerance. It can scale to large groups without placing an undue load on the network, and the message delivery latency scales with the log of the group size. We find that the degree of inconsistency is very low, so that we believe this protocol to be ideal for many wide-area applications.

8.8. ACKNOWLEDGMENTS

John Wilkes, of the Storage Systems Program at Hewlett-Packard Laboratories, and Kim Taylor, of Sybase, assisted the initial development of these protocols. George Neville-Neil and Mary Long gave helpful comments on this work.

8.9. REFERENCES

1. D. AGRAWAL and A. MALPANI, "Efficient dissemination of information in computer networks," <u>Comp. Journal</u>, **34**, 6, pp. 534–41 (December 1991).
2. N. ALON, A. BARAK, and U. MANBER, "On disseminating information reliably without broadcasting," <u>Proc. 7th Int. Conf. on Distributed Computing Systems</u> (1987), pp. 74–81.
3. D. BARBARÁ, and H. GARCIA-MOLINA, "The case for controlled inconsistency in replicated data," <u>Proc. Workshop on the Management of Replicated Data</u> (November 1990), pp. 35–8.
4. A. DEMERS, D. GREENE, C. HAUSER, W. IRISH, J. LARSON, S. SCHENKER, H. STURGIS, D. SWINEHART, and D. TERRY, "Epidemic algorithms for replicated database maintenance," <u>Operating Systems Review</u>, **22**, 1, pp. 8–32 (January 1988).
5. A. EMTAGE and P. DEUTSCH, "Archie – an electronic directory service for the Internet," <u>Proc. Winter 1992 Usenix Conf.</u>, pp. 93–110.
6. R. GOLDING, "End-to-end performance prediction for the Internet," Tech. rep. UCSC–CRL–92–26, Computer and Information Sciences Board, University of California at Santa Cruz (June 1992).
7. R. GOLDING, "Weak-consistency group communication and membership," Ph.D. dissertation, published as Tech. Rep. UCSC–CRL–92–52, Computer and Information Sciences Board, University of California at Santa Cruz (December 1992).
8. R. GOLDING and D. LONG, "Design choices for weak-consistency group communication," Tech. rep. UCSC–CRL–92–45, Computer and Information Sciences Board, University of California at Santa Cruz (September 1992).
9. R. GOLDING and K. TAYLOR, "Group membership in the epidemic style," Tech. rep. UCSC–CRL–92–13, Computer and Information Sciences Board, University of California at Santa Cruz (April 1992).
10. R. LADIN, B. LISKOV, and L. SHRIRA, "Lazy replication: exploiting the semantics of distributed services," <u>Operating Systems Review</u>, **25**, 1, pp. 49–55 (January 1991).
11. L. LAMPORT, "Time, clocks, and the ordering of events in a distributed system," <u>Comm. ACM</u>, **21**, 7, pp. 558–65 (1978).

12. D. LONG, J. CARROLL, and C. PARK, "A study of the reliability of Internet sites," Proc. 10th IEEE Symp. on Rel. Distr. Sys. (September 1991), pp. 177–86.
13. C. PU and A. LEFF, "Replica control in distributed systems: an asynchronous approach," Tech. rep. CUCS–053–090, Department of Computer Science, Columbia University (January 1991).

CHAPTER 9

MODELING A MULTIMEDIA SYSTEM FOR VOD SERVICES

Antonio Puliafito, Salvatore Riccobene,
Giancarlo Iannizzotto, Lorenzo Vita

9.1. INTRODUCTION

The constant increase in the performance and economic accessibiliy of modern calculation and communication systems has directed the interest of both researchers and industry towards the implementation of new forms of highly interactive multimedia communication such as Video on Demand. These systems belong to the Continuous Media (CM) category[1], which feature a continuous flow of data from a service manager (mass storage system) to a multimedia user interface (client).

The implementation of CM systems, however, shows up certain important limits in current mass memory technology, which is essentially electromechanical. For some time now, in fact, disks have constituted serious bottlenecks for performance in certain applications. The recent spread of parallel processing systems and multimedia applications has increased these limits even further.

Besides the inherent technological limitations, one of the main limits of mass storage systems derives from the fact that up to now time constraints have not been considered as high-priority requirements. Designers have mainly concentrated on forms of organization that are capable of ensuring a good trade-off between reliability and average performance. The insertion of a new user (new connection) thus leads to a general reduction in the QoS already connected users are offered.

VOD systems[2], on the other hand, require the transfer of large (and often fixed) amounts of data with strong time constraints (deadlines): a management policy which, in case of insertion of a new user, degrades the QoS of previously connected users is definitely unacceptable. The system has to guarantee each user the bandwidth he requires for the application he is using, from the moment he is inserted to the end of his worksession, with no kind of degradation, unless it is due to accidental events.

A possible solution to the problem of mass memories is given by the parallelization of I/O subsystems by using Disk Array techniques[3]. Disk Arrays are a set of hard disks organized in such a way that the various I/O requests can be served either by different units (parallelization of requests) or by several units at the same time (parallelization of service). The current trend is to organize them in such a way that both kinds of parallelization are possible: in any case, the main goal consists of the uniform distribution of the workload over all the units of the system.

The aim of this chapter is to define and analyze a disk array system for the storage of specific data for CM applications. Focusing on VOD service systems in particular, a redundancy scheme based on an Information Dispersal Algorithm (IDA)[4] is used. A Petri Net model is developed to analyze performance, assessing the overhead introduced and the distribution of response times for a generic read request.

The chapter is organized as follows: Section 9.2 introduces the problem of storing data for VOD applications, describes the solution proposed and outlines the main features of the IDA coding used. Sections 9.3 and 9.4 present the Petri Net model of the system and the results obtained. The authors' conclusions and suggestions for future research are gives in Section 9.5.

9.2. DISK ARRAY FOR MULTIMEDIA SERVERS

As compared with traditional applications, VOD services enjoy a series of specific features, mainly connected with typical properties of the data treated:

MODELING A MULTIMEDIA SYSTEM FOR VOD SERVICES 189

- the transfer rate required is particularly high;

- the data flow has to be kept more or less continuous and is subject to very strong time constraints, unlike common applications;

- as the data transferred is mainly audio/video sequences, it is highly likely that a request for data concerning a certain set of video frames will be followed by a request for the frames which come immediately afterwards. It is therefore reasonable to schedule these requests in anticipation by means of simple prefetching algorithms, thus optimizing and reducing service times;

- the system writing phase takes place off-line or at times when the load is low, so requests subject to time constraints are almost exclusively read requests.

A system designed for the management of VOD services therefore has to be based on an architecture which is able to exploit these particular features.

One possible architecture[2], divided into three logical levels, is composed as follows: the lowest logical level comprises high-capacity mass storage devices with a low access speed and a low cost/storage ratio. This level deals with long-term storage of film sequences.

The central level is the real server of the system and comprises devices which have to offer not only a high storage capacity, but also considerable access speed. These devices also have to be able to manage a large number of connections at the same time. It is thus reasonable to envisage the use of disk array technology.

The third level refers to the end user or user group with particular features in common and provides an access interface with extremely high-speed but low-capacity memory devices (something similar to cache memories) for the temporary storage of data.

In this chapter we will concentrate exclusively on the middle level, describing an architecture based on disk-array technology designed specifically for VOD services.

The main feature of the organization proposed is the adoption of IDA coding[4] in the data storage phase (a more detailed description of IDA is given in Appendix A).

Up to now the main advantage of IDA coding was its potential for enhancing system reliability. The possibility of inserting an arbitrary quantity of redundancy in data treatment, in terms of both storage and transmission, provides high MTBF values for any system. However, with particular reference to mass memory storage systems it should be pointed out that the use of this technique has not been very successful so far. This is surely due to the fact that it is not possible to reconcile the features of IDA with the needs of a general-purpose disk array.

One major drawback lies in the fact that the probability of faults occurring simultaneously in more than one unit is not really very high. With RAID5[3] technology, as spare disks are available, a faulty disk can be replaced very quickly, thus reducing to a minimum the time window during which a fault in a second unit would necessarily cause failure of the whole system. In most applications using a single additional unit to record is sufficient and economically feasible.

Another drawback of using IDA is that there is a certain degradation in performance in write operations, as each operation requires access to all the units in the group. Operations concerning a limited amount of data are therefore greatly penalized.

In this chapter we want to show that IDA coding can be used efficiently with disk array systems and performance can even be enhanced if the workload applied has certain characteristics. The use of a particular management policy related to the typical features of VOD systems, in fact, makes it possible to eliminate the drawbacks outlined above, to the advantage of performance.

The load generated by systems for the management of VOD services consists almost exclusively of read requests, each of which refers to large amounts of data. Write operations, on the other hand, are sporadic and generally take place off-line. In addition, as for read requests, they refer to large amounts of data. This means that the performance drawbacks described above - due to the size of the blocks IDA operates on - are no longer a problem.

IDA coding provides a large number of units from which the same data can be read; a read operation can thus be performed on the fastest units and the system's response times are considerably reduced. The choice does not necessarily have to be made a priori, but can be run-time.

Let us assume, for instance, a system with N disks. They do not all need to be used to store the same file. If, in fact, the value of N is high, it is more convenient to store a file on a subset, called a group, within which IDA coding is used. If G is its size, we get $G = R + B$, where R is the redundancy introduced and B is the base to which this redundancy is applied. If we set $G = 6$ units, for example, and we want a 50% redundancy we will have $B = 4$ and $R = 2$. The files will therefore be subdivided in series of four blocks. Coding is applied to each block, thus obtaining the 6 blocks to be stored on the six disks in the group. Thanks to the IDA technique any subset of B disks can be chosen from the G making up the group to reconstruct the original information. It is clear that as the value of R increases, the number of sets data can be read from increases considerably. The idea is therefore to send a read request to all the disks in the group and wait for the B fastest ones, thus preventing overloaded disks from slowing the system down. For the remaining $G - B$ disks (which are R), there are two possibilities:

1. the corresponding subrequest is being served; in this case, even though an abort operation is still possible, there will be a functioning overhead (the disk will be working unnecessarily);

2. the subrequest is still queued. In this case it can be removed without uselessly loading the relative disk.

The aim of the following section is therefore to determine how much useless work is done by the system disks if such a management policy is used.

9.3. MODEL OF THE SYSTEM

In this paragraph we develop a model of the disk array proposed in order to evaluate the overhead introduced by the management policy used.

Of the various available methods we chose modeling by Stochastic Reward Nets (SRNs)[5], which associate the various markings with reward indexes, so as to obtain effective evaluation indexes. The system examined comprises a set of 12 disks in groups of three units each. The division of the disks into groups was made on the basis of the Partial Dynamic Declustering (PDD) policy described in [6]. The division is not a static one: groups are formed dynamically, according to the current workload conditions on the various units. The main advantage of dynamic management is efficient distribution of the workload over the mass storage subsystem.

The use of IDA coding does not entail any significant change in the PDD technique; indeed, it is quite transparent to it. Fig. 9.1 shows the model of a generic group comprising three units. For the IDA coding we fixed $N = 3$ and $R = 1$, so $B = 2$. A generic read request addressed to the group is split into three subrequests, each addressed to all the disks. The request can be considered to have been served when two units have terminated their part. The model will also have to show the behaviour of the third (and last) subrequest, once the other two have been served. In other words, it is necessary to assess whether it is still queueing or already being served. As the disks also belong to other groups, in fact, as provided for by PPD, the corresponding subrequests proceed asynchronously up the various queues, so both situations are possible. To model this behaviour, the queue for a generic disk has been described using a set of places. Each disk has two types of traffic - one relating to the group being examined and the other to all the other groups it belongs to. Below we will indicate the former as internal and the latter as external.

External requests follow the flow T_ext, Pa, ta, Pb, T_Serv_a. The queue for these requests is subdivided into two places, Pa and Pb. If there are no internal requests queued, ta is enabled and the queued external requests are accumulated in Pb, to be served by the transition T_Serv_a. The arrival of an internal request (firing of TG) places a mark in Pc. If there are other requests already queued (in Pb) tb is blocked, thus preventing the mark from passing to Pd. A mark in Pc disables the transition ta, so all the external requests following the arrival of the internal request will remain in Pa. With this technique it is therefore possible to discriminate between external requests which have arrived before an internal one and those which arrived afterwards. FIFO management of the queue is thus guaranteed, in accordance with the protocol used. Once all the previous external requests have been served, the internal request can be served as well (enabling of tb). We point out that there are three exit routes for the marks representing internal requests: the transitions tc, T_over and T_Serv_b. Trasition tc is enabled when there are two marks in $Pout$. This situation indicates that the minimum number of responses on the part of the disks has been reached. If the third subrequest, the delayed one, is still in Pc, i.e. still queueing (as there are other requests ahead of it), tc eliminates it,

MODELING A MULTIMEDIA SYSTEM FOR VOD SERVICES

thus without it representing a workload for the disk, by using a reward function which returns 1 on the occurrence of the following condition:
$$(\#Pout + \#Pc1 + \#Pc2 + \#Pc3) = 3$$
It can easily be seen that when this condition occurs the system is doing useless work.

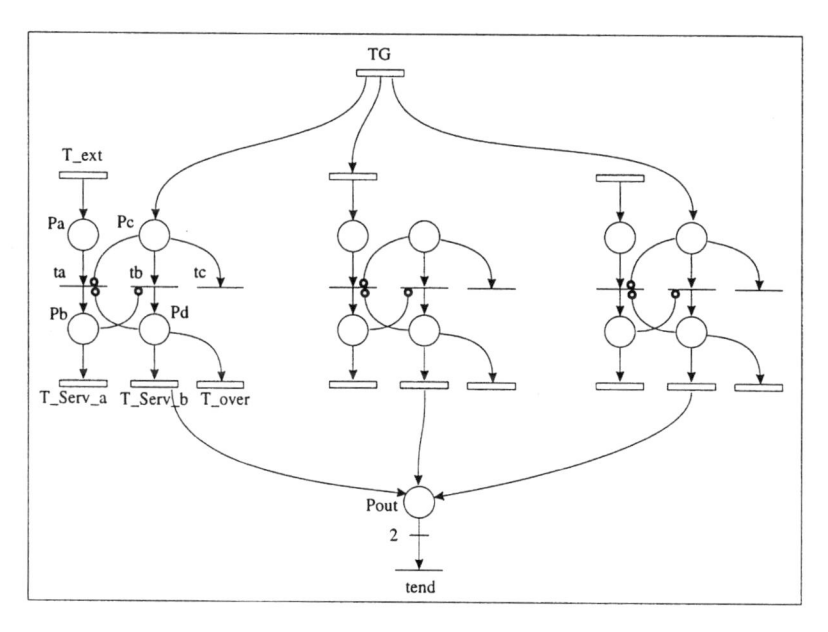

Figure 9.1: Petri Net model.

As regards the characteristics of the disks, the following values are assumed (for each disk):

Substained Transfer rate	5 MBytes/s;
Mean Seek Time	9 ms;
Mean Latency Rotational Time	5 ms.

Table 9.1.: Disk parameters

The workload applied to the system was hypothesized as being made up of multimedia transactions for VOD services, with films coded using the MPEG1 standard. The average throughput per connection is therefore 1.5 Mb/s. Assuming that each request addressed to the system refers to 25KB of data, each connection will involve a workload of 7.5 req/s. This value determines the firing rate of the transition TG.

The rate of T_ext was made to vary from a minimum of 7.5 to a maximum of 37.5, which correspond respectively to a total system workload of 180-540 req/s and therefore 4.5-13.5 MB/s (see table 9.2).

Fig. 9.2 shows the results obtained from an analytical solution of the model. As can be seen, the overhead introduced tends to decrease as the workload increases. This can be accounted for by the fact that, due to the dynamic management of the groups, when the traffic increases the time the corresponding subrequests spend queueing becomes mutually uncorrelated, and so the probability that, once the minimum number of requests needed to reconstruct data has been reached, the other subrequests (only one in our model) will still be queueing ant not in service phase. It can also be seen from the fig. 9.2 that, although the overhead is quite high when workload values are low, it is not much higher than 5% for high values. If low values of workload are assumed the system can support the overhead introduced by this management policy quite well.

On the other hand, the advantages obtained in terms of service times are considerable, as will be pointed out in the following section.

	rate TG (req/s)	rate T_ext (req/s)	Total rate to the system
1	7.5	7.5	180
2	7.5	15	270
3	7.5	22.5	360
4	7.5	30	450
5	7.5	37.5	540

Table 9.2.: Analytical evaluation parameters

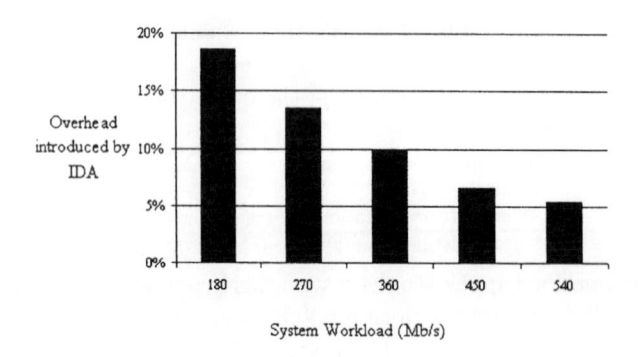

Figure 9.2: Overhead introduced by IDA.

9.4. EVALUATION OF SYSTEM RESPONSE TIME

The aim of this section is to evaluate the architecture described above in strictly performance terms. The parameter we used to do so was the system response time to a generic I/O request.

Response time is identified in literature as the time between generation of a request and completion of the operation requested. In the specific case of mass memory subsystems, it can be divided into six parts:

- Tos = time taken by the O.S. to process the request and translate it into low-level orders;

- Tw = time spent in queueing (in the controller or in the O.S. memory);

- RPS = rotational positioning sensing delay: time to make a full revolution when, in the absence of disk cache, the channel is not free when the disk is ready;

- Ts = seek time: time required for the heads to find the right cylinder;

- Trl = rotational latency time: time necessary to find the required sector;

- Trw = reading/writing time: time required to transfer data to or from the main memory.

In our assessment we will not take the first term into account, as it depends almost exclusively on the implementation of the specific operating system. We will not take the RPS into account either, as it is possible to use techniques to reduce this delay to zero.

The last three terms are generally indicated as the service time, as they mainly depend on the characteristics of the disk. The probabilistic distribution of these times was studied in [7].

According to [7] we model the disk access delay, the sum of the seek and latency times, with a normal distribution and the transfer time with a deterministic time depending on the requests. To deal with non exponentially distributed events, we use phase type approximation[8]. The generic transition T_Serv_x, which models the service time in Fig. 9.1 is thus replaced by two cascades of Erlang distributions, the characteristics of which are given in Table 9.3.

Distribution	Order	Mixing Probability	Mean [ms]	Single Stage Rate [ms-1]
1	10	0.2926	1.455	6.869
2	10	0.7073	27.106	0.369

Table 9.3.: Parameters for the two mixed Erlang distributions

For a more detailed description of the procedure see [9].

To calculate the response time *pdf*, we used the tagged customer method. We calculated the steady-state probability of the states of the equivalent Markov chain as seen by a generic customer. Then the $MTTA$ was calculated for each state, considering the absorbing state to be the one in which the tagged customer has been served. This method is described in detail in [5].

Obviously in our specific case the tagged customer coincides with the arrival of an internal request. The network in Fig. 9.1 was modified as shown in Fig. 9.3. To calculate the steady-state probabilities, the part relating to each disk has been simplified by combining places Pa and Pb in a single place $Pext$. The network in Fig. 9.3b, on the other hand, represents the model to be solved to obtain the $MTTA$ values. In this case the place $Pout$ becomes the absorbing one.

The results obtained using this method are given in Fig. 9.4, which shows the probability distribution functions for applied load values of 9, 11.25 and 13.5 MB/s.

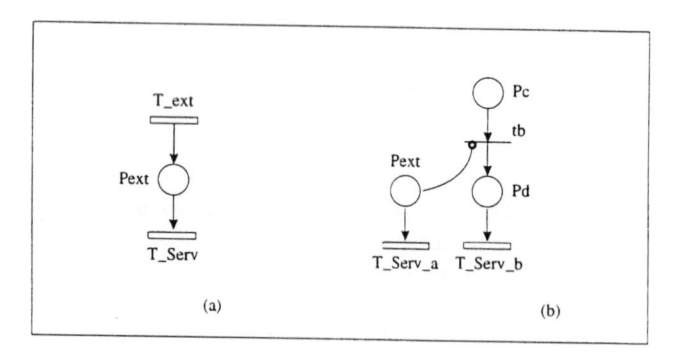

Figure 9.3: Petri Nets for the Tagged Customer Method.

MODELING A MULTIMEDIA SYSTEM FOR VOD SERVICES 197

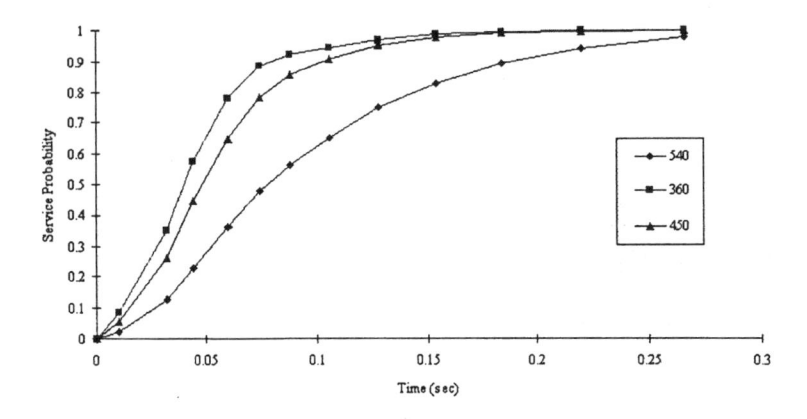

Figure 9.4: Response time distributions.

As can be seen, even with very low deadlines (100, 150 ms) the system has a very high probability of serving requests. This means that in a VOD service the probability of loss of videoframes is very low. Even when the workload is high, the system manages to serve the requests in a short time (200 ms). Knowledge of these response time probability distribution curves is indispensable for the design of a VOD system as it allows an appropriate size to be chosen for the Disk Array and a maximum threshold to be determined for the acceptable number of connections.

9.5. CONCLUSIONS

With the organization presented in the chapter it is possible to implement an efficient server for VOD services. The Petri net model described allows the performance obtainable from the system to be evaluated. The work also shows the great advantages that can be had, in terms of performance, from combined use of Partial Dynamic Declustering and IDA coding in the presence of typical CM workloads. Further improvements could be obtained by managing the service queues with prefetching algorithms, as it is possible to predict future requests on the basis of those currently being served. This allows high-quality multimedia services to be provided.

9.A. APPENDIX - THE IDA TECHNIQUE

The *IDA* technique functions as follows. In an initial phase the information is subdivided in such a way as to form fixed-size macroblocks which contain the minimum unit processed by *IDA* in each operation. Appropriate choice of the size of these macroblocks therefore has to be made according to the features of the system and the data to be processed.

Each macroblock is then divided further into M blocks of a constant size, D (see Fig. 9.5).

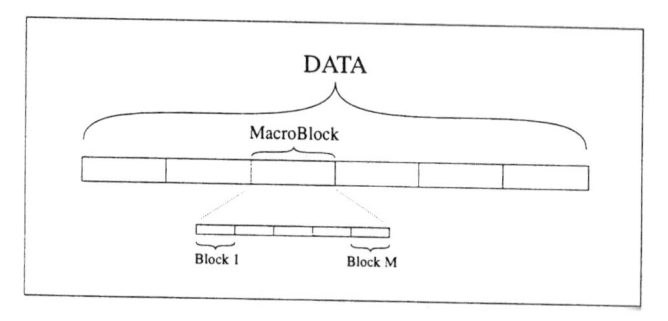

Figure 9.5: Ida coding.

If R is the desired redundancy value, i.e. R represents the number of units to be added, then the total number of units needed for storage will be $N = M + R$. Starting from the M initial blocks, the algorithm will build N new blocks, also of size D (see Fig. 9.6). Each of the N output blocks will be obtained by a linear combination of the M input blocks. To perform this operation it is necessary to use an $M \times N$ matrix. Choice of the matrix (i.e. the coding) is such that any minimum subset of M elements chosen from the N output blocks will be made up of linearly independent elements. In other terms, starting from any subset of M elements from the N outputs, it is possible to reconstruct the M input blocks. The only condition required for decoding is identification of the blocks from which to decode the original information, i.e. the index (from 1 to N) associated to each element.

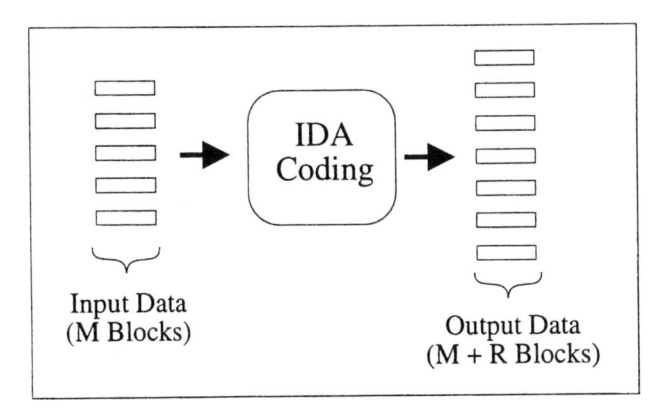

MODELING A MULTIMEDIA SYSTEM FOR VOD SERVICES 199

Figure 9.6: *IDA* coding phase.

Parity encoding can be seen as a particular case of *IDA* in which the first M blocks of the N to be recorded are exactly the same as the initial data and the $M+1-th$ block is calculated by an XOR operation on the input data. With parity encoding we always have $R = 1$ and $N = M + 1$. Normally *IDA* is used to obtain an increase in the $MTBF$ of a system. According to the system's reliability features, it is possible to choose the value, R, of redundancy in such a way as to obtain the total $MTBF$ value desired. Even if the system has a very low $MTBF$ value, suitable choice of the value of R will bring it back to satisfactory values. Table 9.4 gives the $MTTF$ values for a disk array system with a space of 10 disks. It compares solutions based on $RAID1$, $RAID5$ and *IDA*. the disks are all the same and have an $MTBF$ of 1000h. The basic system (without redundancy) as a whole will have an $MTBF$ of 1000h (0.11 years).

Scheme	Disks no.	MTBF
None	11	0.11
RAID 5	11	23
IDA	11	23
IDA	12	5284
IDA	13	1393837
IDA	20	3.7×10^{15}
RAID 1	20	114

(values expressed in years)
Table 9.4: MTTF values for a disk Array with B=10 disk units

These values were calculated according to [3] and do not take into account situations of correlated disk failures, which would considerably reduce the values given above. As regards the MTBF for RAID1, only the case corresponding to 20 disks is reported, because in this case total duplication of the disks is required (every disk has a mirror disk). Similarly, for the RAID5 case only the value corresponding to 11 disks is given, because only one disk is required to store parity information. The table also highlights the great flexibility of the IDA solution for redundancy purposes.

REFERENCES

1. D.J. Gemmel, H.M. Vin, D.D. Kandlur, PV. Rangan, L.A. Rowe, 'Multimedia storage servers: a tutorial", Computer, Vol. 28, No 5, May 1995.

2. T.D.C. Little, D. Venkatesh, "Prospects for interactive Video-on-Demand", IEEE Multimedia, Vol. 1, No 3, Fall 1994.
   This paper is a survey on the technological considerations for designing a large-scale, distributed, interactive multimedia system. It examines the basic problems and proposes some implementation solutions.

3. P.M. Chen, E.K. Lee, G.A.Gibson ,R.H. Katz, D.A. Patterson , "RAID: High-Performance, Reliable Secondary Storage", ACM Computing Surveys, Vol. 26, No 2, June 1994.
   This work is a detailed survey dealing with the problematics of Disk Arrays and standardized architectures. In particular, a detailed description of RAID systems is presented.

4. M.O. Rabin, "Efficient Dispersal of Information for Security Load Balancing and Fault Tolerance", ACM Journal of the Association for Computing Machinery, Vol.36 No 2, April 1989, pp 335-348.
   This paper describes a new information coding system which can be used for storing and for transmitting data on a network not totally reliable.

5. J.K. Muppala. K.S. Trivedi, V. Mainkar, V.G. Kulkarni, "Numerical computation of response time distributions using stochastic reward

nets", Annals of Op. Research, 48 (1994), pp.155-184.

6. V. Catania, A. Puliafito, S. Riccobene, L. Vita, "Design and Performance Analysis of a Disk Array System", IEEE Transaction on Computer, Vol. 44, No 10, October 1995, pp. 1236-1247.
This work presents a new technique, named Partial Dynamic Declustering, for dynamic management of disk groups in a Disk Array. Comparisons are made with other existing architectures.

7. M.Y. Kim, A.N. Tantawi, "Asynchronized disk interleaving: Approximating access delays", IEEE Transaction on Computer, Vol.40, No. 7, July 1991, pp. 801-810.

8. M.A. Johnson, "Selecting Parameters of phase distributions: combining nonlinear programming, heuristics and Erlang distributions", ORSA J. on Computing, Vol.5, No. 1, 1993, pp. 69-83.

9. M. Malhotra, A.L. Reibman, "Selecting and implementing phase approximations for semi-Markov models", Stochastic Models, 9 (4), pp. 473-506, 1993.

CHAPTER 10

MODELING, SIMULATION, AND SYNTHESIS OF HIGH-PERFORMANCE ATM PROTOCOLS AND MULTIMEDIA SYSTEMS

Georg Carle and Jochen Schiller

10.1. INTRODUCTION

Emerging applications mostly require both high performance as well as support for a wide variety of communication services. For example, audio, video, and data transmission may require highly different services, e.g., guaranteed delay, jitter, or bandwidth. An additional challenge arises through the growing demand for multipoint communication services. ATM networks are capable of satisfying the basic application requirements by providing multipoint bearer services[6] with data rates exceeding a gigabit per second. However, current communication subsystems (including higher layer protocols) that provide reliable services are not able to deliver the available network performance to the applications.[14,35] In particular in multipoint communication scenarios, severe degradations of service quality can be observed. Additional problems need to be addressed in scenarios where quality of service (QoS) requirements and processing capabilities of individual receivers differ.

In order to provide the required high performance services to the applications, new protocols[29,25] as well as high-performance implementation architectures for the communication subsystems need to be designed.[28,16,5] Dedicated VLSI components should be used in flexible implementation platforms for time-critical processing tasks, such as retransmission support or memory management, in order to provide high performance communiation services.

This chapter presents a new approach for the flexible design of hardware-supported high-performance communication subsystems together with a framework for the provision of multipoint multimedia services in ATM networks and heterogeneous internetworks. The design process allows mapping of a formal protocol specification onto a parallel, hardware-based implementation architecture. The highly modular VLSI implementation architecture designed with parameterizable and programmable components allows for service flexibility. The architecture is not limited to a certain protocol, but allows the implementation of a variety of high-speed protocols. We validated our approach with a design example using a formal specification of the protocol RMC-AAL (Reliable Multicast ATM Adaptation Layer, RMC-AAL[9]). This protocol provides error control functions for ATM end and intermediate systems in order to enhance the reliability of an ATM service. The concept of integrating error control functions into Group Communication Servers (GCSs) allows for the efficient provision of reliable multipoint services in large, widespread groups. The multicast error control capabilities of GCSs allow for increased throughput and reduced delay. GCSs provide protocol processing support for multicast transmitters and reduce the acknowledgement implosion problem.[10] They also support groups consisting of end systems with direct ATM access, as well as end systems connected over heterogeneous internetworks.[8]

This chapter is structured as follows. Section 2 presents the developed ATM protocol for multipoint error control. Section 3 describes the design flow for implementation in detail and gives some examples for each design step. Section 4 presents conclusions and potential further work.

10.2. ATM PROTOCOL FOR MULTIPOINT ERROR CONTROL

The Reliable Multicast ATM Adaptation Layer features the options of frame-based automatic repeat request (ARQ), cell-based ARQ and forward error correction (FEC) for an efficient provision of reliable multicast services under varying cell loss rates. RMC-AAL offers a fully reliable service and a service that assures delivery to a subset of K receivers. It can be used in ATM end systems and also in dedicated servers within the network (see Figure 10.1).

Lost retransmissions contribute significantly to the QoS of a reliable group communication service.[10] It is therefore of high importance to decrease the probability of lost retransmissions. This goal can be achieved by FEC.[3] The mean number of retransmissions required for the successful delivery of a frame can also be decreased by establishing virtual channels (VCs) for retransmissions which have a lower cell loss rate than the VCs used for the first transmission of a frame. This capability of ATM is in contrast to

DESIGN OF HIGH-PERFORMANCE ATM SYSTEMS 205

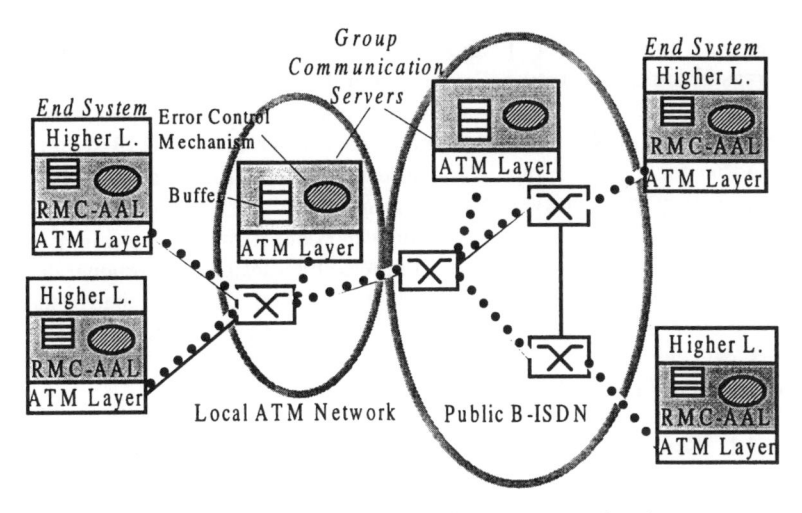

Figure 10.1: Multicast error control in servers and end system

single service networks, where initial transmissions and retransmissions will generally observe identical loss rates.

RMC-AAL allows to send retransmissions by multicast or by unicast in selective repeat or go-back-N mode. It can be selected if retransmissions are frame-based (by retransmission of the original data frames) or cell-based (by retransmission of frame fragments). When FEC is used, the information cells of the frame are protected by additional redundancy cells. Encoding and decoding can be based on Reed-Solomon-Codes,[23] or on simple XOR-operations and matrix interleaving.[27]

In the following, the data format used by RMC-AAL is briefly explained. RMC-AAL consists of a service specific convergence sublayer (SSCS) with ARQ and FEC functions, based on the common part convergence sublayer (CPCS) of AAL5.[17] AAL5 uses a trailer of 8 bytes and protects the payload of an AAL frame by the cyclic redundancy check CRC-32. In addition to this 8 byte trailer, RMC-AAL data frames use a 10 byte frame header. Frames are identified by a frame sequence number (FSN, 24 bit) in the frame header. When using cell-based ARQ, each cell has an additional protocol overhead of one byte: 2 bits for specifying the cell type, and a 6 bit cell sequence number (CSN). Even for high speed VCs in WANs, no large cell numbering space is required, because a hierarchical sequence numbering scheme is used, where each cell is identified by both FSN and CSN. The alternative solution of identifying cells entirely by their cell sequence numbers leads to a significantly higher overhead per cell. For example, the protocol BLINKBLT,[13] which also offers cell-based retransmissions, has a per-cell overhead of 4 bytes. The RMC-AAL frame header contains a transmitter identifier and the length of the SSCS PDU payload. The frame header also contains a discriminator field with an identifier for the frame type, a flag to request an immediate acknowledgement, a flag for identifying the last frame of a burst, and a field for the number of redundancy cells that follow the data frame. Frame fragments consist of a Fragment Header Cell, followed by a selection of original data cells of this frame. The fragment header cell contains the header information of a regular frame, as well as a bitmap for identification of the data cells which follow. Further details of the RMC-AAL protocol are presented in Carle and Zitterbart.[9]

10.2.1. Group Communication Server

The application of the presented error control mechanisms is not limited to ATM end systems. The deployment of so-called Group Communication Servers with multicast error control mechanisms provides reliable high-performance multipoint services for a wide range of parameters. Further improvements of performance and efficiency can be achieved by using GCSs hierarchically.

GCSs support an efficient use of network resources by performing multicast error control within the network. Allowing retransmissions originating from the server avoids unnecessary retransmissions over common branches of a multicast tree. The integration of FEC mechanisms into the GCS allows for the regeneration of lost cells and for the reinsertion of additional redundancy for adjusting the FEC coding scheme according to the needs of subsequent hops.

By providing protocol processing support for multicast transmitters, GCSs also improve scalability. They also support heterogeneous multicasting by allowing different protocol parameters for different branches of a multicast tree, and by converting between different error schemes. For example, the simple frame-based go-back-N retransmission scheme could be used between a GCS and local receivers, while the more complex cell-based ARQ scheme with additional FEC could be used for a wide area connection between transmitter and GCS.

For groups with multiple transmitters, the GCS provides support for multiplexing of frames onto a single point-to-multipoint connection. This reduces the number of required VCs significantly for large groups with many transmitters.[36] Virtual LANs frequently require this multiplexing functionality. However, an additional queuing delay may be introduced by

DESIGN OF HIGH-PERFORMANCE ATM SYSTEMS 207

this multiplexing function. If LAN Emulation[1] is used in a local ATM network, a GCS might be incorporated into the service elements 'LAN Emulation Server' (LES) and 'Broadcast and Unknown Server' (BUS), thus making it possible for applications to ensure the reliable delivery of multicast and broadcast messages to all peers.

10.2.2. Formal Protocol Specification

RMC-AAL for end systems as well as for GCSs has been formally specified in SDL (Specification and Description Language).[18] Figure 10.2 shows the structure of a simplified GCS. Each octagon represents a process in SDL.

A data frame arriving from the sender enters the diagram at the upper right corner and is lead to the process Frame Manager Receive (FM_Receive) by the receiver process Filter_snd, which forwards frames depending on the frame type. FM_Receive allocates memory and stores the frame. The frame is then scheduled for transmission by the process Send_Manager. The process Frame Manager Send (FM_Send) assembles the head of the frame and passes the frame to the transmitter process (Switch_rcv). This process ensures that cells of different frames are not interleaved. The Pool_Manager manages the buffer of the GCS and the lower window edges (LWEs) of all receiving entities, which can be either end systems, or GCSs. Acknowledgements arrive at the process Filter_rcv. FM_Repeat interprets the acknowledgements and passes the results on to the Pool_Manager. The core of the frame-based error handling is formed by the process Frame_Control. Acknowledgement frames are created by the process FAck_Creator.

10.2.3. Protocol Simulation

It is important to know which error control scheme of RMC-AAL is best suited for a given situation. The achievable performance of the proposed error control schemes was evaluated by applying discrete-event simulation with the simulation tool BONeS/Designer.[31] For modeling the correlation properties of lost cells, a two state Markov chain (Gilbert Model) was applied. Based on the worst case observations of Ohta and Kitami,[24] a probability of 0.3 was used for a cell discard following a cell discard. This is equivalent to cell losses with a mean burst length of 1.428 cells. A multicast VC was simulated, where cell losses may occur on the common link or on individual links. The same error model was applied to all links. A data rate of 100 Mbit/s, a distance of 100 km, and a frame length of 50 cells was used. Figure 10.3 shows the efficiency (relation of successfully

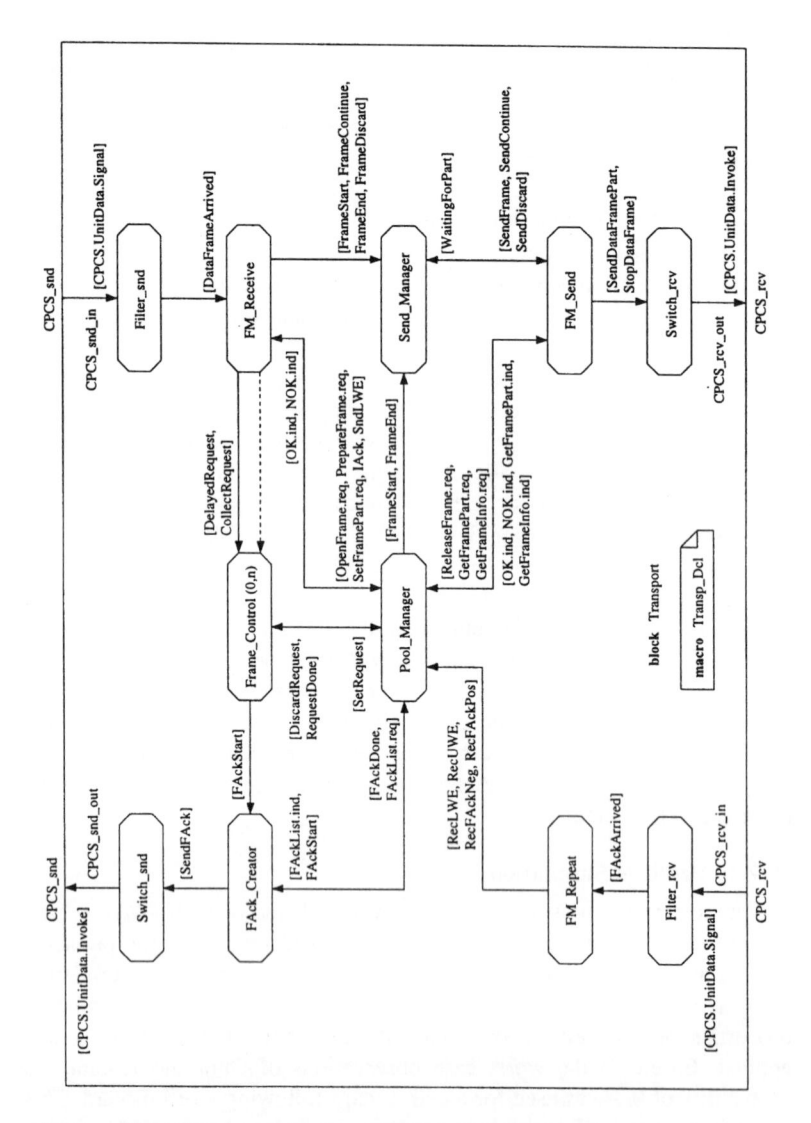

Figure 10.2: SDL specification of RMC-AAL for Group Communication Servers

DESIGN OF HIGH-PERFORMANCE ATM SYSTEMS 209

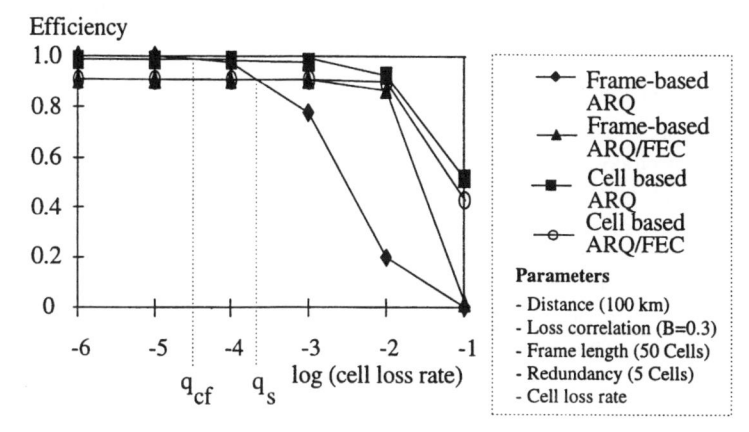

Figure 10.3: Simulation of efficiency for different error control schemes

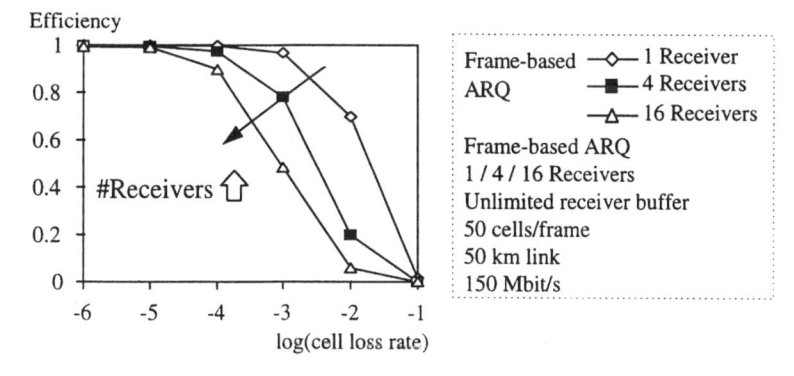

Figure 10.4: Influence of a growing number of receivers

transmitted useful frames to total number of transmitted frames) for the four schemes frame-based ARQ, frame-based ARQ/FEC, cell-based ARQ, and cell-based ARQ/FEC for varying cell loss probability.

Figure 10.4 shows how the achievable efficiency of frame-based ARQ decreases for an increasing number of receivers.[8]

In order to select an appropriate error control mechanism, the following question is of interest: up to which cell loss probability does a frame-based ARQ scheme result in higher efficiency than a framebased hybrid ARQ scheme? An interpolation of the simulation results shows a cell loss probability q_s of approximately $\log(q_s) = -3.4$. A similar threshold q_{cf} exists for the efficiency equilibrium of the cell-based and the frame-based

ARQ. An analytical treatment of these questions can be found in G. Carle.[10]

10.3. IMPLEMENTATION

For the provision of high performance communication services, not only suitable protocols but also high performance implementations are required. Traditional implementation by hand is error prone due to the high concurrency of different processes forming the protocol. Therefore, it is much better to derive an implementation of a protocol at least semi-automatic. But this is only practicable, if the synthesis tools are powerful enough to produce implementations that fulfill the performance requirements. The following section shows some design steps performed with commercial tools in combination with tools we developed ourselves, resulting in high-performance implementations.

10.3.1. Design Flow

A significant amount of research analyzes the automatic derivation of a high-performance communication subsystem from a formal specification.[22,20] Figure 10.5 shows this goal embedded into several additional steps, representing the design flow of CHIMPSY (communication-oriented high-performance modular processing system).[26] The *specification* mentioned here consists of the *protocol* itself and a set of *configuration parameters*. These parameters comprise the chosen integration alternative, technology, and interface depending on the desired performance, the existing software environment, and other non-formal values. From these parameters an *implementation framework* is derived.[4]

The configuration parameters describe the desired performance, the existing software environment, and the maximum costs of a system. Costs may be expressed in terms of processing costs or hardware complexity. We use simulation and measurement results of architectures synthesized in previous design cycles to determine the required number of processing units.

Using these parameters, an *implementation framework* is composed from a set of predefined functional units. This framework consists of the interfaces to an environment, static memory for protocol data, and a central crossbar to connect all components (see Figure 10.6). We describe all hardware components shown in Figure 10.6 using the standardized hardware description language VHDL (VHSIC Hardware Description Language[15]).

DESIGN OF HIGH-PERFORMANCE ATM SYSTEMS 211

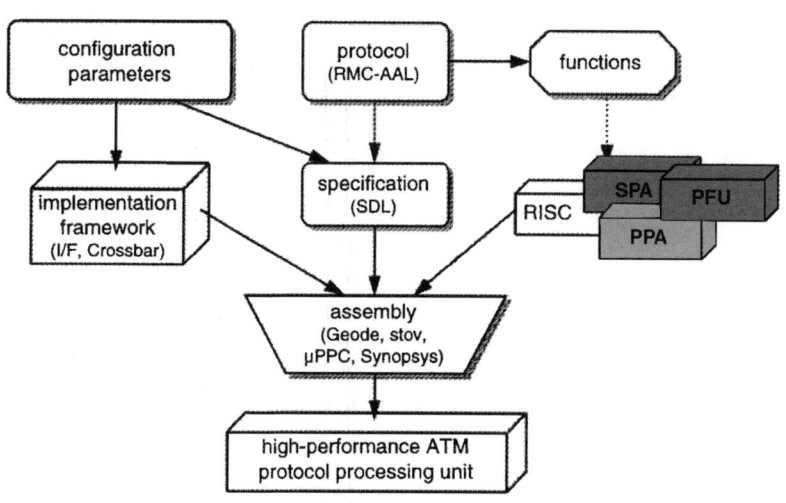

Figure 10.5: Customized design flow for high-performance ATM protocol processing units

10.3.2. Architectures for Implementation of Functional Units

From *protocols* like RMC-AAL we have extracted several *functions*, e.g., timer, CRC, FEC, transmit, and acknowledgement processing. These functions can be implemented on four different alternative architectures depending on the desired performance:

- *RISC-processors*: The tool GEODE[33] can be used to generate C-code from an SDL-specification. A RISC-processor can execute this code after compilation. Descriptions of different RISC-processors are available for example as VHDL-code or gate-level schematic.

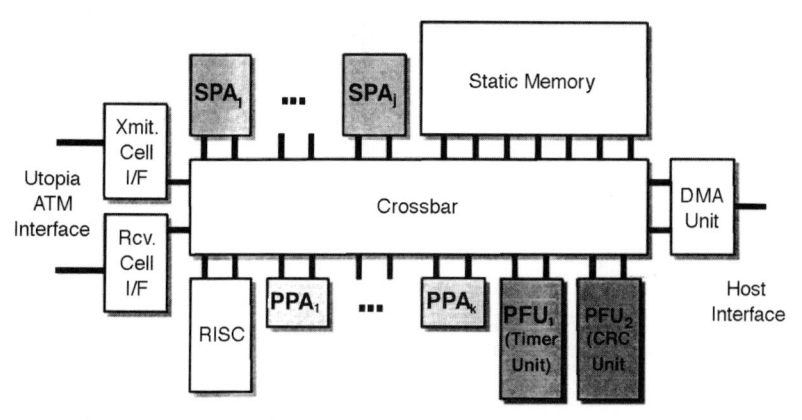

Figure 10.6: Flexible architecture for high-performance ATM protocol processing

- *Synthesizable Protocol Automata (SPA)*: With the help of a customized SDL-to-VHDL-compiler we can automatically map an SDL-description of a protocol automaton onto a VHDL-description of a hardware unit. After synthesis onto real hardware (e.g. with the tool Synopsys[30]) this unit acts as a protocol automaton. Due to the dedicated hardware for protocol processing, the performance of such a unit may be significantly higher compared to a general purpose RISC-processor. However, existing hardware synthesis tools do not achieve optimal performance when synthesizing netlists from high-level VHDL descriptions.
- *Programmable Protocol Automata (PPA)*: For even higher performance we have designed microprogrammable automata. These automata consist of only 2895 standard cells in CMOS-technology and can run with 100 MHz. Up to now, we program these units directly with microcode using a custom microcode compiler µPPC (microprogram protocol compiler).
- *Protocol Function Units (PFU)*: To achieve highest performance, we implement time-critical protocol functions as hardware macros. Examples are timer, CRC, and FEC units. Gate-level VHDL is used to implement PFUs.

Depending on the specification, the different units are chosen and configured to assemble the *high-performance protocol processing unit*. Currently, we are using 0.7µm standard-cell technology for layout synthesis and, alternatively, FPGA-boards inserted into workstations for rapid prototyping.

10.3.3. Design Tools

Our design flow comprises several tools as shown in Figure 10.5. Up to now it is not possible to find a single tool for the whole design flow that is flexible enough for the different requirements and that produces communication components with the required performance. The following items give a short overview of the tools used in our approach, and summarizes their advantages and disadvantages.

- *SDL-to-C compiler (GEODE code generator)*: GEODE[33] is a commercial tool set for the design of event-driven real time systems, using the language SDL'88,[18] and Message Sequence Charts (MSC[19]) for formal protocol specification. The tool set provides support for graphical editing, simulation, debugging, and C-code generation. Both the graphical form of SDL called SDL/GR and the textual phrase form called SDL/PR are supported. SDL specifications are logically composed of a hierarchy of structural objects. It can be selected how the

DESIGN OF HIGH-PERFORMANCE ATM SYSTEMS

GEODE code generator maps the SDL objects process, process instance, block, and system onto operating system processes. Specific functionalities which are specified as abstract data types can be mapped onto separately specified C functions. In our flexible design approach, we also map abstract data types onto specific hardware functions implemented in Protocol Function Units.

- *SDL-TO-VHDL compiler (stov)*: In order to facilitate the process of hardware implementation of SDL specifications we developed a dedicated SDL-TO-VHDL compiler called *stov*. The compiler generates VHDL code that is adapted for the flexible architecture shown in Figure 10.6. The generated code makes use of the existing VHDL libraries that describe the architecture. This allows for rapid prototyping of protocol processing units after successful simulation of the SDL specification. As there are some SDL constructs that cannot be translated into hardware descriptions, an appropriate subset of SDL is supported by the compiler.
- *VHDL compiler (Synopsys)*: Based on VHDL-descriptions of hardware on the register-transfer level, this commercial tool synthesizes netlists for different technologies.[30] These netlists can be used for further synthesis on ASICs or FPGAs. Compared to hand-coded netlists, this tool does not achieve the optimum speed and size of hardware due to the complexity of the synthesis. On the other hand, using such a powerful synthesis system is the only way to manage the complexity of large hardware systems. In addition to synthesis, this tool also allows for simulation and debugging of VHDL-descriptions.
- *Microcode compiler (μPPC)*: Our custom microcode compiler allows for easy programming of the PPAs using a simple assembly level language. The language comprises 19 operations, comments, labels, and macros.

The compiler converts this microcode into a binary format which can be downloaded to the PPA. The disadvantage of this microcode is its low level language. Therefore, an additional SDL-to-Microcode-Compiler is under development.

10.3.4. Protocol Implementation

Several non time-critical finite state machines (FSM) of a protocol can be mapped onto a single PPA in the implementation architecture. Table 10.1 shows some lines from the specification of a retransmission FSM. The table comprises the actual *state*, incoming *events*, and *conditions* to check before *actions* and the transition into the *next state* are performed.

State	Event	Conditions	Actions	Next State
NULL	i.new_context		init_context(ARR)	ACTIVE
ACTIVE	i.rec_closing		signal(TI, i.rec_seq)	ACTIVE
GAPTEST		gap_no = 0	s.base = s.offset ...	ACTIVE

Table 10.1: Example lines from the retransmission FSM

The shaded line shows that the FSM issues the signal *i.rec_seq* to the FSM *TI* in the state *ACTIVE* as soon as the event *i.rec_closing* has been received. The next state is also *ACTIVE*. The translation of this line into microcode for the implemented FSM is shown in Table 10.2. The first three lines wait for receiving a signal via the input port and branch to the appropriate piece of microcode after receiving the signal. The last five lines send the signal *i.rec_seq* to the FSM *TI*, save the new state (*ACTIVE*), and jump back to the beginning to wait for the next signal.

10.3.5. Synthesis

As one example for a resulting component, Figure 10.7 shows the layout of a PPA synthesized using Synopsys[30] and Cadence[7] together with the 0.7 µm CMOS standard-cell library from ES2.[11] Due to a critical path of 9.6 ns of the control logic this chip can run with approximately 100 MHz. This means that every 10ns one row of microcode from Table 10.2 can be executed. We implemented together with the control logic 11 kbyte of SRAM, with 9.5 kbyte for microcode (approx. 1750 rows) and the remaining 1.5 kbyte for registers and stacks. The chip area is 52.5 mm² with a utilization of 78%. This size is mainly due to the SRAM technology used with the prototype design.

DESIGN OF HIGH-PERFORMANCE ATM SYSTEMS

data	selection	command	control	comment
		CNT	GETH	wait for signal
`move in,base`		CNT	MOVE S,A	load context
		BRV	MOVE H1,A	jump to subroutine
`1`		CNT	MOVE S,C	signal
`TI;` `i.rec_seq`		CNT	SGNL	i_rec_seq to TI
`OFFSET;` `move mem,out`		CNT	MOVE A,Q	
`ACTIVE`		CNT	SAVE	next state
`label`		JMP	SLEEP	return

Table 10.2: Microcode example of a FSM suited for a PPA

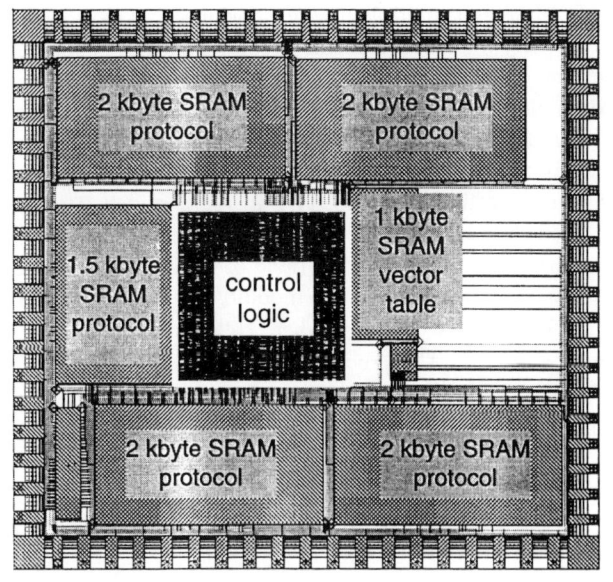

Figure 10.7: VLSI chip layout of the programmable protocol automata

10.3.6. Performance Evaluation

In order to study processing delay and implementation complexity, the prototype implementation of a GCS on a network adapter with the following properties was investigated: Protocol processing is performed on one or more 32 bit RISC processors with an average performance of 100 MIPS. In addition, hardware support for segmentation and reassembly, hardware for CRC32 calculation, and hardware for FEC processing were assumed. Based on a specification of the processes of a GCS in assembly-level language, the number of instruction cycles necessary to perform the required functionality was determined for each process of the GCS. Subsequently, the delay of a process that is executed on a processor with 100 MIPS was evaluated.

In Figure 10.8, the processing delays are compared to the cell interarrival time of 2.74 µs for an ATM link of 155 Mbit/s. The figure shows the processing delays for GCS configurations of increasing complexity from left to right. The configuration with the lowest processing delays is a GCS with frame-based ARQ. The figure also shows the additional processing delays of a GCS with cell-based ARQ, and with additional FEC. The right edges of the bars indicate the processing delays in each component if cell-based ARQ, FEC, and multiplexing are performed. The figure illustrates several properties. First, the receiver process and the transmitter process have a constant delay independent of any ARQ or FEC processing. Second, frame-level multiplexing does not take much time in any component other than the Send Manager which does the scheduling of the frames. Note that the GCS does not provide a copy function for multicasting. This copy function for multicasting is provided by an ATM switch. Third, and most important, it can be seen that the delay is dominated by the Frame Manager Receive whenever the first cell of a frame of a connection with cell-based ARQ and FEC is processed. In this module, the processing delay of the first cell of a frame is 2.71 µs when cell-based ARQ is selected in combination with FEC. The processing delay of a cell in the middle of a frame is 1.31 µs, while the processing delay of the last cell of a frame is 1.47 µs. Thus, this component is the first candidate for optimization, and for the deployment of hardware components for processing support.

A single processor of 100 MIPS leads to a processing bottleneck at high loads, as the overall delay for processing of a cell by the GCS (summarizing the processing times of all modules) is larger than the cell interarrival time. A set of three processors with support of dedicated hardware to perform table lookup, filtering, and the construction of outgoing cells allows for maximum load with an ATM link of 155 Mbit/s even for frames consisting of a single cell. Not shown in the diagram are queuing delays of cells that have to wait because frames of other senders in the same group have to be sent first. Furthermore, operations caused by the processing of acknowledgements in a GCS or in a sending host are not

DESIGN OF HIGH-PERFORMANCE ATM SYSTEMS 217

contained in the diagram. The latter operations heavily depend on the number of receivers that acknowledge the reception of frames or cells, and on the acknowledgement strategy (e.g., NAKs might be sent as soon as possible after detection of an error, as opposed to ACKs which are sent cumulatively).

This performance evaluation shows clearly that for higher performance dedicated VLSI has to be used. With the assumed RISC processors and only moderate hardware support only a bandwidth of 155 Mbit/s can be achieved. For implementation in intermediate systems one has to use the above described hardware components.

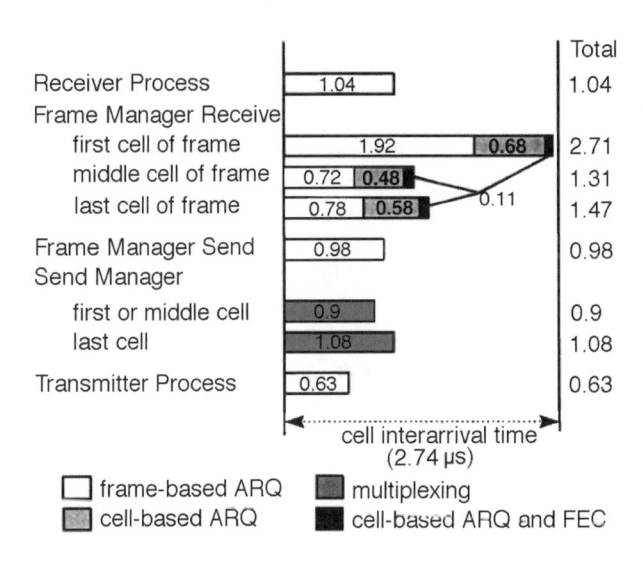

		Total
Receiver Process	1.04	1.04
Frame Manager Receive		
first cell of frame	1.92 / 0.68	2.71
middle cell of frame	0.72 / 0.48	1.31
last cell of frame	0.78 / 0.58 / 0.11	1.47
Frame Manager Send	0.98	0.98
Send Manager		
first or middle cell	0.9	0.9
last cell	1.08	1.08
Transmitter Process	0.63	0.63

cell interarrival time (2.74 µs)

☐ frame-based ARQ   ■ multiplexing
▨ cell-based ARQ   ■ cell-based ARQ and FEC

10.3.7. Summary of the Design Flow for Protocol Implementation

Figure 10.9 shows a simplified diagram of the design flow used for the hardware implementation of high-performance communication protocols. Cycles in the design flow have been omitted for clarity. The two basic technologies used in our design are CMOS ASICs and FPGAs. Therefore, the figure shows the synthesis and derivation of parameters for these two technologies in more detail. Furthermore, we use FPGA-boards inserted in standard workstations for rapid prototyping.[34]

10.4. CONCLUSION & FURTHER WORK

A new framework for multipoint error control is presented which has the potential to fulfill many requirements. For small groups and low cell loss rates, a frame-based end-to-end error control is most appropriate. Cell-based retransmission as well as FEC allow high-performance reliable multicasting even for significant cell loss rates. For better scalability and support of heterogeneous scenarios, the deployment of a new network element called the Group Communication Server (GCS) is proposed. It allows an hierarchical approach for multicast error control and the combination of different error control schemes.

Investigation of the processing delay demonstrated the feasibility of the proposed error control schemes even for very high speeds. It also revealed that cell-based error control schemes contribute little to the processing load for error-free transmissions.

A basic idea of our work is the semi-automatic implementation of high-performance protocols based on formal specifications. This design process allows for rapid prototyping, and gaining valuable insight into the tradeoff of protocol performance and protocol processing costs. With this design flow, even processing-intensive cell-based algorithms can be implemented at high speeds, using a high-performance parallel protocol architecture.

Based on communication requirements, e.g. group communication in the context of multimedia applications and ATM networks, we develop communication protocols and describe them using formal description techniques. One example for this is the standardized language SDL. Together with a generic implementation architecture we can now derive a specific implementation. The generic implementation architecture consists of several basic components with different flexibility and performance. Depending on the protocol and performance requirements these basic components are combined and configured. We have developed several tools to support the different design steps, e.g. the mapping of SDL specifications onto VHDL hardware description. In addition, for the last design steps towards real hardware implementation we use commercial design tools.

Future work will concentrate on the design of more complex units to support protocol processing and the implementation of these units in workstations as prototypes. Furthermore, additional design tools for high-performance communication subsystems are under development.

DESIGN OF HIGH-PERFORMANCE ATM SYSTEMS 219

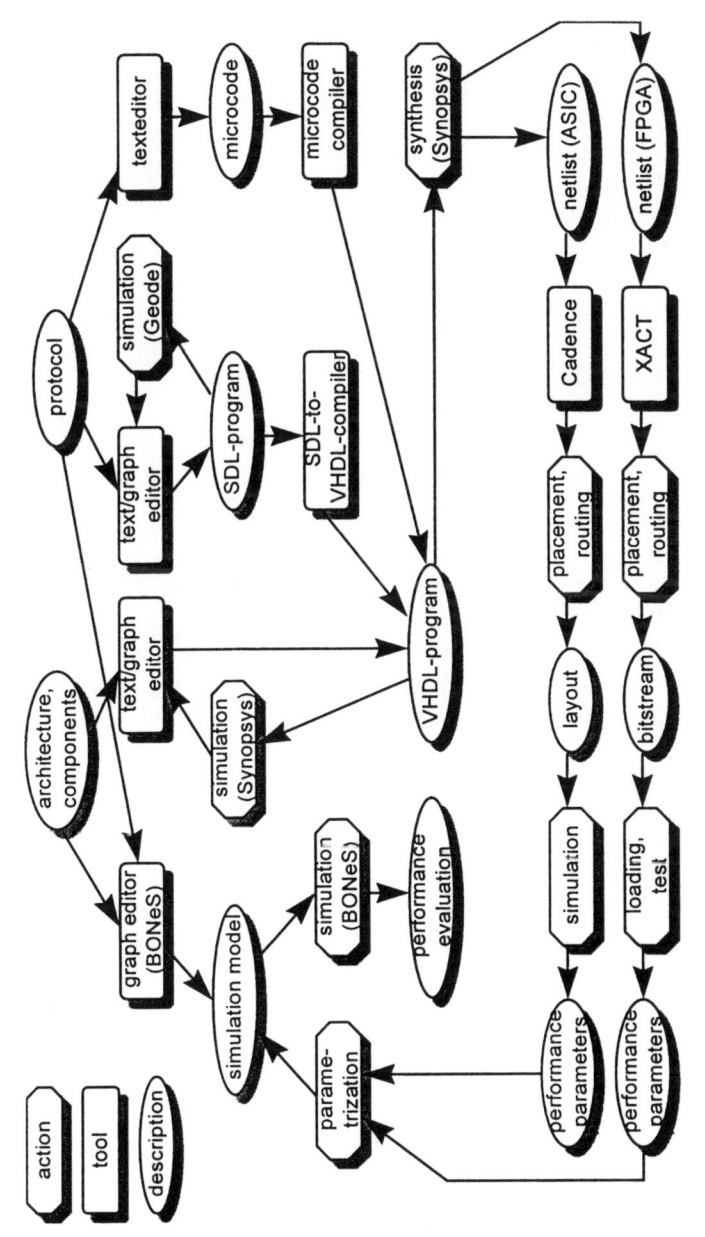

Figure 10.9: Simplified implementation design flow

REFERENCES

1. ATM FORUM, LAN Emulation Over ATM: Draft Specification, LAN Emulation Sub-working Group, ATM Forum Technical Committee, August 1995
2. BALRAJ, T.; YEMINI, Y., „Putting the Transport Layer on VLSI - the PROMPT Protocol Chip," in: Pehrson, B.; Gunningberg, P.; Pink, S. (eds.): Protocols for High-Speed Networks, III, 1992, North-Holland, pp. 19-34
3. BIERSACK, E. W., „Performance Evaluation of Forward Error Correction in an ATM Environment," IEEE Journal on Selected Areas in Communication, 11, 4, pp. 631-640, (1993)
4. BRAUN, T.; SCHILLER, J.; ZITTERBART, M., „A Highly Modular VLSI Implementation Architecture for Parallel Transport Protocols," IFIP 4th International Workshop on Protocols for High-Speed Networks, Vancouver, Canada, August 1994
5. BRAUN, T.; ZITTERBART, M., „Parallel Transport System Design," in: Danthine, A.; Spaniol, O. (eds.): High Performance Networking, IV, IFIP, North-Holland, 1993, pp. 397-412
6. BUBENIK, R.; GADDIS, M.; DEHART, J., „Communicating with virtual paths and virtual channels," Proceedings of the 11th INFOCOM'92, pp. 1035 - 1042, Florence, Italy, May 1992
7. CADENCE DESIGN SYSTEMS, INC., Documentation for DFW II (Design Framework II), Cadence Design Systems, Inc., San Jose, California, 1994
8. CARLE, G., SCHILLER, J., „Enabling High-Bandwidth Applications by High-Performance Multicast Transfer Protocol Processing," 6th IFIP Conference on Performance of Computer Networks, Istanbul, Turkey, October 23-26, 1995, in S. Fdida, R. Onvural (Eds.): Data Communications and their Performance, Chapman&Hall 1996, pp. 82-96
9. CARLE, G., ZITTERBART, M., „ATM Adaptation Layer and Group Communication Servers for High-Performance Multipoint Services," 7th IEEE Workshop on Local and Metropolitan Area Networks, pp. 98-106, March 26-29, 1995, Duck Key, Marathon, Florida, USA
10. CARLE, G., „Towards Scalable Error Control for Reliable Multicast Services in ATM Networks," 12th International Conference on Computer Communication, ICCC'95, Seoul, Korea, August 20-25, 1995

11. EUROPEAN SILICON STRUCTURES, Documentation for 0.7μm-Library, European Silicon Structures, Rousset, France
12. FELDMEIER, D.C., „An Overview of the TP++ Transport Protocol," in: Tantawy A.N. (ed.): High Performance Communication, (Kluwer Academic Publishers, 1994)
13. GOLDSTEIN, F., „Compatibility of BLINKBLT with the ATM Adaptation Layer," ANSI Technical Subcommitee T1S1.5/90-009, Raleigh, NC, USA, Feb. 1990
14. HEINRICHS, B.; JAKOBS, K.; CARONE, A., „High performance transfer services to support multimedia group communications," Computer Communications, 16, 9, (1993)
15. IEEE, Standard VHDL Language Reference Manual, IEEE Std 1076-1987
16. ITO, M.; TAKEUCHI, L.; NEUFELD, G., „Evaluation of a Multiprocessing Approach for OSI Protocol Processing," Proceedings of the First International Conference on Computer Communications and Networks, San Diego, CA, USA, June 1992
17. ITU-T, Recommendation I.363, BISDN ATM Adaptation Layer (AAL) Specification, Geneva, 1993
18. ITU-T, Recommendation Z.100: Functional Specification and Description Language (SDL), Telecommunication Standardization Sector of ITU, Geneva, 1988
19. ITU-T, Recommendation Z.120: Message Sequence Chart (MSC), Telecommunication Standardization Sector of ITU, Geneva, 1993
20. KRISHNAKUMAR, A.S., „A Synthesis System for Communication Protocols," Proceedings of the 5th Annual IEEE International ASIC Conference and Exhibit, Rochester, New York, September 1992
21. KRISHNAKUMAR, A.S.; KNEUER, J.G.; SHAW, A.J., HIPOD, „An Architecture for High-Speed Protocol Implementations," in: Danthine, A.; Spaniol, O. (eds.): High Performance Networking, IV, IFIP, (North-Holland, 1993), pp. 383-396
22. KRISHNAKUMAR, A.S.; KRISHNAMURTHY B.; SABNANI, K., „Translation of Formal Protocol Specifications to VLSI Designs," Protocol Specification, Testing and Verification, VII, Elsevier Science Publishers B.V., (North-Holland, 1987), pp. 375-390
23. MCAULEY, A., „Reliable Broadband Communication Using a Burst Erasure Correcting Code," ACM SIGCOMM '90, Philadelphia, PA, USA., Sep. 1990

24. OHTA, H.; KITAMI, T., „A Cell Loss Recovery Method Using FEC in ATM Networks," IEEE Journal on Selected Areas in Communications, 9, 9, (1991), pp.1471-1483
25. SANTOSO, H.; FDIDA, S., „Transport Layer Multicast: An Enhancement for XTP Bucket Error Control," in: Danthine, A.; Spaniol, O. (eds.): High Performance Networking, IV, IFIP, (North-Holland, 1993)
26. SCHILLER, J., „CHIMPSY - a Modular Processor-System for High-Performance Communication," 1. GI/SI Jahrestagung, Zurich, September 1995
27. SHACHAM, N.; MCKENNY, P., „Packet recovery in high-speed networks using coding," in Proceedings of IEEE INFOCOM '90, San Francisco, CA, pp. 124-131, June 1990
28. STERBENZ, J.P.G.; PARULKAR, G.M., „AXON Host-Network Interface Architecture for Gigabit Communications," in: Johnson, M. J. (ed.): Protocols for High-Speed Networks, II, (North-Holland, 1991), pp. 211-236
29. STRAYER, W.T.; DEMPSEY, B.J.; WEAVER, A.C., XTP: The Xpress Transfer Protocol, (Addison-Wesley, 1992
30. SYNOPSYS INC., Documentation of Simulator, Design Compiler, and Design Analyzer, Version 3.2a, Synopsys, Inc., Mountain View, California, USA, 1995
31. THE ALTA GROUP, Block Oriented Network Simulator™ (BONeS™) User's Guide, The Alta Group of Cadence Design Systems, Inc., USA, 1994
32. THE XTP FORUM, XTP Protocol Definition Proposed Revision 4.0, 1994
33. VERILOG SA, Technical documentation of the GEODE toolset, Verilog SA, Toulouse, France
34. VIRTUAL COMPUTER CORPORATION, EVC1s Technical Reference, Virtual Computer Corporation, Reseda, Kalifornien, USA, May 1995
35. WATERS, A. G., „Multicast Provision for High Speed Networks," 4th IFIP Conference on High Performance Networking HPN'92, Liège, Belgium, December 1992
36. WEI, L.; LIAW, F.; ESTRIN, D.; ROMANOW, A., LYON, T.: „Analysis of a Resequencer Model for Multicast over ATM Networks," 3rd International Workshop on Network and Operating Systems Support for Digital Audio and Video, San Diego, CA, USA., 1992

CHAPTER 11

MODELING TECHNIQUES FOR PCS NETWORKS

Yi-Bing Lin

11.1. INTRODUCTION

A personal communications services (PCS) network[3] is a wireless network that provides communication services for *PCS subscribers*. The service area of a PCS network is partitioned into several sub-areas or *cells*. Each cell is covered by a radio port. The port locates a subscriber or *portable*, and delivers calls to and from the portable by means of paging within the cell it serves. A *registration area* (RA) consists of an aggregation of cells, forming a contiguous geographical region. To connect a phone call to a roaming portable, it is necessary to identify the portable's RA. The strategies commonly proposed are two-level hierarchies[16] that maintain a system of home and visited databases (*Home Location Registers* or HLR, and *Visitor Location Registers* or VLR). To order PCS services, a PCS subscriber must "enroll" with a particular PCS provider. When enrolling, the PCS subscriber gives the PCS provider the necessary information, such as credit, service type, and current location, to setup the PCS account. This PCS account information is stored in the HLR of the PCS provider. When the PCS subscriber roams to another RA, which is likely to be owned by another PCS provider, the PCS subscriber becomes a "visitor" of that RA. The VLR of this RA is used to store the visiting PCS subscriber's information. Upon registering with the VLR, the VLR notifies the

HLR of the visiting PCS subscriber that "your subscriber is at my place".

General models are needed to understand different aspects of large-scale PCS networks (such as user location strategies[8, 13], registration strategies[20, 19, 23], hand-off or automatic link transfer strategies[21, 25, 26], and channel allocation strategies[5, 9, 11, 14, 29]) so that the network will provide a high quality of service to mobile subscribers while minimizing the resource cost incurred by the PCS provider. Two widely studied approaches to modeling are simulation and analytic techniques. The primary advantage of simulation based techniques is that systems can be modeled at any level, giving the system designer a great amount of flexibility. However, this flexibility comes at the price that the simulation model may be too time consuming to use, particularly for models of large-scale PCS networks. On the other hand, analytic techniques offer an alternative solution, but require the development of models that are both accurate (simplifying assumptions regarding the behavior of PCS subscribers are required) and sufficiently simple to be solved. In practice, network designers must have both simulation and analytic models at their disposal to properly evaluate and design PCS networks.

This article is concerned with analytic models for large-scale PCS networks. Analytic models are able to capture the behavior of a system in a concise mathematical formulation without the high computational overheads associated with large simulations. Accordingly, we propose two analytic models that will reduce the effort involved in studying the behavior of large-scale PCS networks. We present two new models for PCS networks that can be solved using analytic techniques. The portable population model is based on the flow equivalent assumption (the rate that portables move into a cell is equal to the rate that they move out of the cell). This model enables determination of the steady-state portable population distribution in a cell that is independent of the portable residence time distribution. The model can be used to study the blocking probability of low (portable) mobility PCS networks, and the performance of portable de-registration strategies. Finally, we describe a model for portable movement. By assuming that the arrival of calls to a portable form a Poisson process, and portable residence times have a general distribution, this model can be used to study location tracking and hand-off algorithms.

The central result of this article is to illustrate that effective, practical models for large-scale PCS networks are available to provide essential quantitative information to network designers.

11.2. THE POPULATION MODEL

We describe a portable population model developed in[14]. An important fact observed from this model is that the steady state portable population in a cell is insensitive to the distribution of the portable residence times.

Let N be the expected number of portables in a cell. Suppose that the residence time of a portable in a cell has a general distribution $F(t)$ with mean $\frac{1}{\eta}$. In the steady state, the rate that portables move into a cell equals to the rate that they move out of the cell. In other words, the rate that portables move in a cell is $\lambda^* = N\eta$. The arrivals of portables can be viewed as being generated from N input streams which have the same general distribution with arrival rate η. The net input stream to a cell can be approximated as a Poisson process with arrival rate $\lambda^* = N\eta$. Thus, the distribution for the portable population can be modeled by an $M/G/\infty$ queue with arrival rate λ^* and the service rate η. Let π_n be the steady state probability that there are n portables in the cell. It can be shown that[10]

$$\pi_n = \left(\frac{\lambda^*}{\eta}\right)^n \frac{e^{-\frac{\lambda^*}{\eta}}}{n!} \qquad (11.1)$$

$$= \frac{N^n e^{-N}}{n!} \qquad (11.2)$$

Figure 11.1 (a) illustrates the population distribution when $N = 50$. The solid curve plots π_n based on Equation (11.2). The "o" marks represent π_n obtained from simulation[18] where the portable residence times are exponentially distributed, and the "$\star$" marks are simulation results for a uniform portable residence time distribution. The figure indicates that the population distribution for a cell is insensitive to these portable residence time distributions and Equation (11.2) is consistent with the simulation results.

Three applications of the population model are listed below:

Low mobility PCS network Equation (11.2) can be used to study the blocking probability p_b for a PCS network of portables with

Figure 11.1: The population distribution

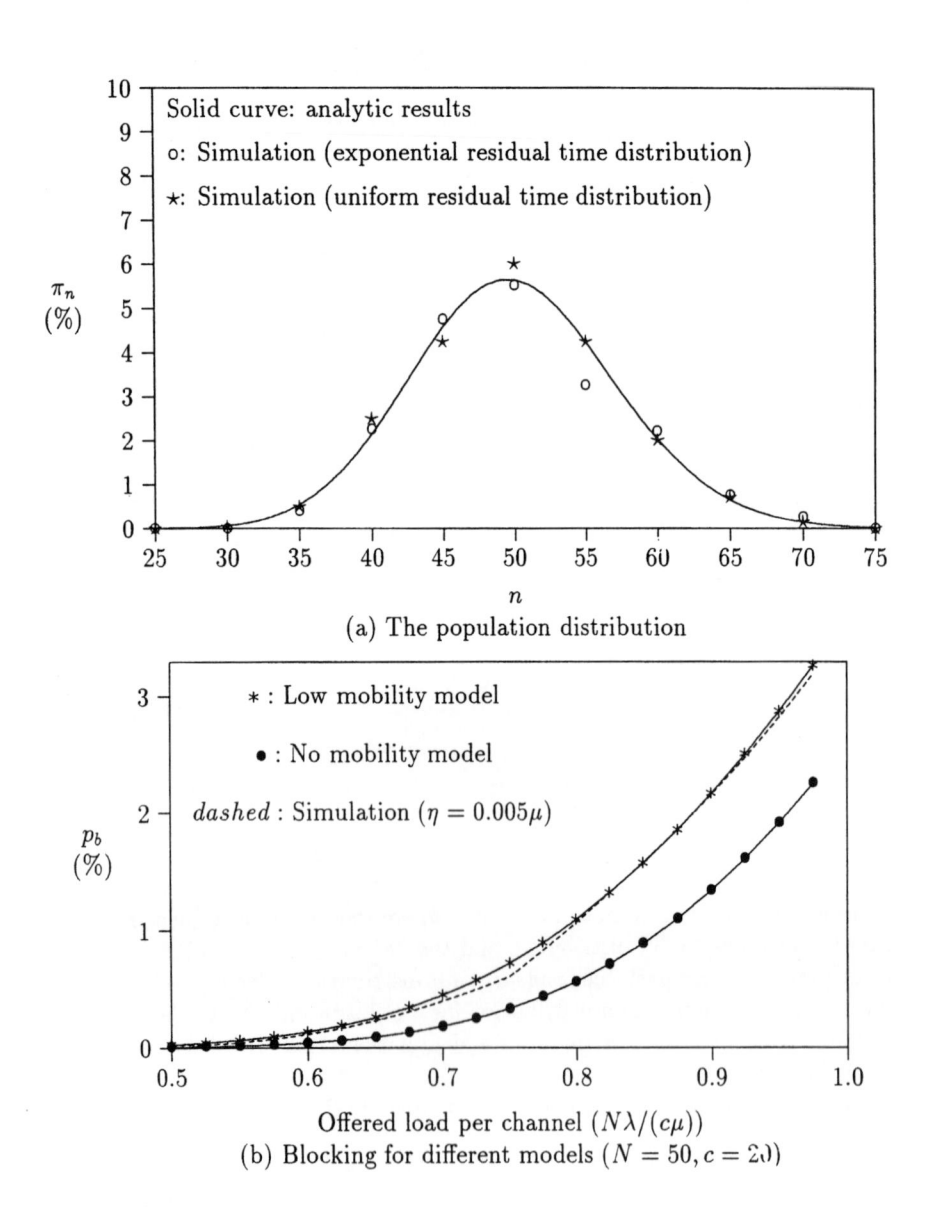

(a) The population distribution

Solid curve: analytic results
o: Simulation (exponential residual time distribution)
★: Simulation (uniform residual time distribution)

π_n (%)

n

(b) Blocking for different models ($N = 50, c = 20$)

∗ : Low mobility model
• : No mobility model
$dashed$: Simulation ($\eta = 0.005\mu$)

p_b (%)

Offered load per channel ($N\lambda/(c\mu)$)

low mobility[14]. The idea is the following. First, the blocking probability $p_b^{(n)}$ is derived for a cell when there are n portables. Then with the low mobility assumption, the blocking probability for the cell is

$$p_b = \sum_{c<n<\infty} \pi_n p_b^{(n)} \qquad (11.3)$$

where c is the number of channels in a cell. Consider a PCS network where every cell has $c = 20$ channels. A phone call is dropped immediately if no channel is available. Suppose that the phone calls to/from a portable are a Poisson process with rate λ, and the holding time is an exponentially distributed random variable with mean $1/\mu$. Figure 11.1 (b) plots the blocking probability p_b against $N\lambda/(c\mu)$, the offered load carried by a channel. The curve marked • represents the the no-mobility model which is a blocking system with finite sources[13]. In this model, the number of portables N in a cell is fixed. When $N \gg c$, the no-mobility model approaches the Erlang-B system. The curve marked ⋆ represents the low-mobility model (i.e., Equation (11.3)) where $p_b^{(n)}$ in (11.3) is derived from the no-mobility model with population n. The dashed curve represents simulation results when the mobility $\eta = 0.005\mu$. The simulation model is described in[18]. Figure 11.1 (b) indicates that our low-mobility model is consistent with the simulation study, while the no-mobility model cannot accurately predict the cell blocking probability appropriately (as already discussed in[15]).

Portable registration In a PCS network, registration is the process by which portables inform the network of their current location (i.e., registration area). A portable registers its location when it is powered on and when it moves between registration areas. The size of a registration database is limited by the amount of resources, such as *temporary local directory numbers*. (Note that it is possible to increase the database size at reasonably low cost to accommodate all portables in the system. However, the amount of resources is limited, and the resources must be reused). We say that the database is full if a portable arrives, and no resources are available for the portable. In this case, the portable cannot access the services provided by the PCS network. When a portable leaves an RA, or shuts off for a long period of time, the portable should be de-registered from the RA, so that any resources previously assigned to the portable can be deallocated.

In IS-41[1, 4], the registration process ensures that a portable registration in a new RA causes de-registration in the previous RA. This approach is referred to as *explicit de-registration*. Bellcore *Wireless Access Communications Systems* (WACS)[2] specifies that a portable should be de-registered by default after a certain time period elapses without the portable re-registering. This scheme is referred to as *timeout de-registration*[23]. In the above registration schemes, it is important to determine (based on the amount of resources for an RA) the probability β that a portable cannot register (and receive service) because no resources are available. In[17, 19], the population model is used to determine β for different registration schemes. Let M be the amount of resources available in an RA. Suppose that the portable residence times have an arbitrary distribution, and let N be the expected number of portables in an RA, then β_{ED} (the probability β for explicit de-registration) is

$$\beta_{ED} = \sum_{M \leq n < \infty} \pi_n \qquad (11.4)$$

where π_n is given in (11.2). For the timeout scheme, the arrivals of portables seen by the scheme is the same as the explicit de-registration. That is $\lambda^* = N\eta$. However, the expected portable residence time $E[\tau]$ seen by the timeout scheme is longer than $\dfrac{1}{\eta}$ because the scheme is only sure that a portable leaves the RA if the portable does not send a re-registration message within a timeout period T. For exponential portable residence time distribution, $E[\tau]$ can be expressed as[17]

$$E[\tau] = \frac{T}{1 - e^{-\eta T}}$$

Thus, β_{TO} (the probability β for timeout de-registration) can be expressed by (11.1) and (11.4) where η in (11.1) is replaced by $1/E[\tau]$; i.e.,

$$\beta_{TO} = \sum_{n=M}^{\infty} \pi_n^* \quad \text{where} \quad \pi_n^* = \left(\frac{N\eta T}{1 - e^{-\eta T}}\right)^n \frac{e^{-\frac{N\eta T}{1 - e^{-\eta T}}}}{n!}$$

PCS simulation In a PCS simulation, portables are assigned for every cell initially. A simple approach is to assign the same number of portables N to each cell. However, it has been shown that this approach results in long simulation execution times. Because in the steady state the distribution of portables in the cells approaches (11.2), we propose generating the initial number of portables in a cell based on (11.2), allowing the PCS simulation to converge more rapidly. This approach will significantly reduce the execution time of the simulation.

11.3. THE PORTABLE MOVEMENT MODEL

Here we describe a portable movement model[12] to study the patterns of the incoming calls and portable movement. The probability $\alpha(K)$ that a portable moves across K RAs between two phone calls is derived assuming that the incoming calls to a portable are a Poisson process, and the time the portable resides in an RA has a general distribution. This study indicates that for a portable with different RA residence time distributions (such as exponential, constant, and uniform distributions) the $\alpha(K)$ distributions are similar if the portable mobility η is high, and the $\alpha(K)$ distributions are very different if the portable mobility is low.

Let t_c be the time interval between two consecutive phone calls to a portable p. Suppose that the portable resides in an RA R_0 when the previous phone call arrived. After the phone call, p visits another K RAs, and p resides in the ith RA for a time period t_i ($0 \leq i \leq K$). Let t_m be the time interval between the arrival of the previous phone call and the time when p moves out of R_0. The relationship among t_c, t_i and t_m is shown in Figure 11.2. We make the following assumptions:

- The phone calls to a portable are a Poisson process. In other words, t_c is exponentially distributed with mean $E[t_c] = 1/\lambda$.

- t_i are independent and identically distributed random variables with a general density function $f(t_i)$, and mean $E[t_i] = 1/\eta$.

The probability $\alpha(K)$ that p moves across K RAs between two phone

Figure 11.2: The relationship among $t_c, t_i,$ and t_m

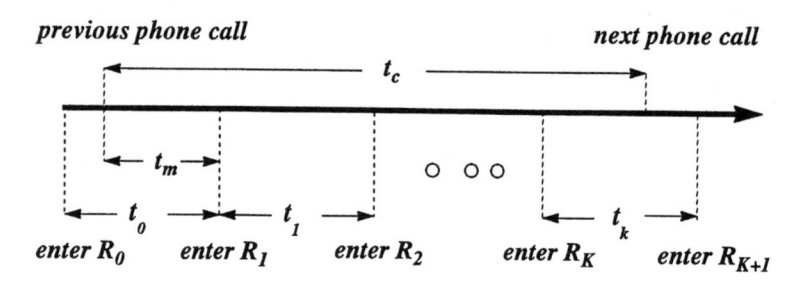

calls is

$$\alpha(K) = \begin{cases} \Pr[t_m + t_1 + ... + t_{K-1} < t_c \leq t_m + t_1 + ... + t_K] & K \geq 1 \\ \Pr[t_c < t_m] & K = 0 \end{cases} \quad (11.5)$$

From the random observer property of the Poisson call arrivals, the distribution of t_m can be derived from $f(t_0)$ using the excess life formula in the renewal theory[24]. Then from the distributions for t_c, t_i and t_m, and Equation (11.5), $\alpha(K)$ is derived[12] as

$$\alpha(K) = \begin{cases} \frac{\eta}{\lambda}\left[1 - f^*(\lambda)\right]^2 \left[f^*(\lambda)\right]^{K-1} & K \geq 1 \\ 1 - \frac{\eta}{\lambda}\left[1 - f^*(\lambda)\right] & K = 0 \end{cases} \quad (11.6)$$

where

$$f^*(s) = \int_{t=0}^{\infty} e^{-st} f(t) dt \quad (11.7)$$

is the Laplace-Stieltjes Transform for $f(t_i)$. Equation (11.6) is general enough to accommodate any portable mobility patterns (i.e., arbitrary $f(t_i)$ functions; note that the Laplace pairs for many functions are already available[22, 27]).

If t_i is exponentially distributed, then from (11.6), $f^*(\lambda) = \frac{\eta}{\eta + \lambda}$, and

$$\alpha(K) = \begin{cases} \dfrac{\eta^K \lambda}{(\eta + \lambda)^{K+1}} & K \geq 1 \\ \dfrac{\lambda}{\eta + \lambda} & K = 0 \end{cases} \quad (11.8)$$

The intuition behind (11.8) is the following. Since t_0 is exponentially distributed, the "excess life" t_m has the same distribution as t_0[24]. Since both t_c and t_m are exponentially distributed,

$$\alpha(0) = E[t_c < t_m] = \int_{t_m=0}^{\infty} \int_{t_c=0}^{t_m} \lambda e^{-\lambda t_c} \eta e^{-\eta t_m} dt_c dt_m = \frac{\lambda}{\eta + \lambda}$$

which is consistent with the case $K = 0$ in (11.8). Now consider $K > 0$. When the portable moves into R_i, the remaining time before the next phone call has the same distribution as t_c (due to the memoryless property of an exponential distribution). If the portable does not receive the next phone call before it enters R_i, then the probability that the portable receives the next phone before it leaves R_i is $q = \frac{\lambda}{\eta + \lambda}$. Thus, $\alpha(K)$ has a geometric distribution

$$\alpha(K) = q(1-q)^K = \left(\frac{\lambda}{\eta + \lambda}\right)\left(\frac{\eta}{\eta + \lambda}\right)^K$$

The result is the same as the case $K \geq 1$ in (11.8).

For the constant portable residence time distribution with mean η, $f^*(\lambda) = e^{-\frac{\lambda}{\eta}}$, and

$$\alpha(K) = \begin{cases} \frac{\eta}{\lambda}\left(1 - e^{-\frac{\lambda}{\eta}}\right)^2 e^{-\frac{(K-1)\lambda}{\eta}} & K \geq 1 \\ 1 - \frac{\eta}{\lambda}\left(1 - e^{-\frac{\lambda}{\eta}}\right) & K = 0 \end{cases}$$

For the uniform portable residence time distribution in the range $\left[0, \frac{2}{\eta}\right]$, $f^*(\lambda) = \frac{\eta}{2\lambda}\left(1 - e^{-\frac{2\lambda}{\eta}}\right)$, and

$$\alpha(K) = \begin{cases} \frac{\eta}{\lambda}\left[1 - \frac{\eta}{2\lambda}\left(1 - e^{-\frac{2\lambda}{\eta}}\right)\right]^2 \left(1 - e^{-\frac{2\lambda}{\eta}}\right)^{K-1} & K \geq 1 \\ 1 - \frac{\eta}{\lambda}\left[1 - \frac{\eta}{2\lambda}\left(1 - e^{-\frac{2\lambda}{\eta}}\right)\right] & K = 0 \end{cases}$$

Figure 11.3 plots $\alpha(K)$ for different portable residence time distributions with different λ/η values. Figures 11.3 (a) and (b) indicate that

Figure 11.3: $\alpha(K)$ for different portable residence time distributions

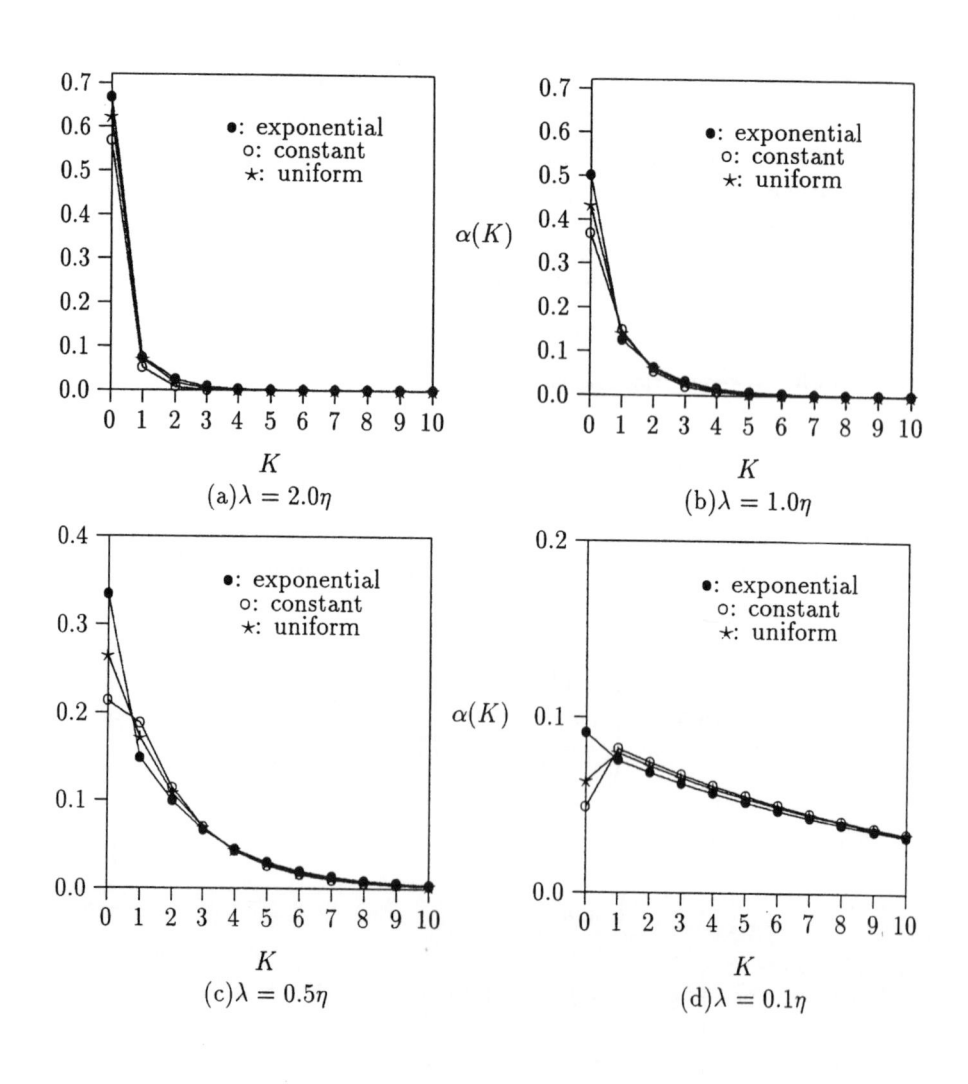

(a) $\lambda = 2.0\eta$

(b) $\lambda = 1.0\eta$

(c) $\lambda = 0.5\eta$

(d) $\lambda = 0.1\eta$

MODELING TECHNIQUES FOR PCS NETWORKS 233

when $\lambda > \eta$, the $\alpha(K)$ distribution is not sensitive to the portable residence time distributions. On the other hand, Figures 11.3 (c) and (d) indicate that when $\lambda < \eta$, for a small K, the probability $\alpha(K)$ is significantly affected by the portable residence time distributions.

Two applications of the portable movement model are described below.

Location Tracking In a PCS system, the RA of a called portable must be determined before the connection can be established. Due to the mobility of portables, the database HLR is required to store the location (i.e., address of the visited RA) of a portable. The location record is modified when the portable moves to another RA. Every RA is associated with a VLR. (An VLR may serve one or more RAs. For the demonstration purposes, we assume that every VLR serves an RA.) In IS-41, when a portable moves from an RA to another RA, it registers at the VLR of the new RA, and its new location is reported to the HLR.

A technique called *location forwarding* was proposed to reduce the location update cost in a PCS network[8]. The idea is described as follows. When a phone moves to a new RA, no message is sent to update the HLR. Instead, a message is sent to the old RA to create a forwarding pointer to the new RA. The cost of creating a forwarding pointer is assumed to be cheaper than the modification of the location record at the HLR. When an incoming call arrives, the forwarding pointers are traced to find the actual location of the phone.

Suppose that during two incoming calls, the phone moves to a new RA K times, and the number of forwarding pointers traced to find the actual location is k. Then the cost saved in the location forwarding (compared with the IS-41 scheme) is K operations to update the HLR. On the other hand, the extra penalty paid in the location forwarding is k operations to trace the forwarding pointers when a portable is located. Since the phone may revisit an RA, we have $k \leq K$. Thus, the key issue of modeling location forwarding is to derive the values for k and K. The K distribution can be derived based on the portable movement model (i.e., (11.5)). The derivation of k can be done using a two-dimensional random walk with reflecting barriers[12].

Hand-off Algorithms As noted earlier when a portable moves from one cell to another while a call is in progress, the call requires a new channel (in the new cell) to continue. If no channel is available in the new cell, then the call will be dropped or forced terminated. The forced termination probability is an important criterion in the performance evaluation of the PCS network. Forced termination of an ongoing call is considered less desirable than blocking of a new call attempt. Several hand-off schemes have been proposed (see[25] for a survey) to reduce forced termination. Performance modeling of the hand-off schemes were intensively studied. In most modeling efforts (either analytic analysis or simulation)[6, 7, 21, 25, 28], a single cell is studied by assuming that both new call attempts and the hand-off calls are Poisson processes with arrival rate λ_n and λ_h, respectively. Then modeling techniques are used to simulate the behavior of a specific hand-off scheme with aggregative call arrivals where output measures such as p_o (the new call blocking probability), p_f (the forced termination probability), and p_{nc} (the probability that a call cannot complete due to blocking or forced termination) are derived. In this model, λ_h cannot be arbitrary selected. Rather, λ_h, p_o and p_f are affected by each other and are affected by the portable mobility rate η. Based on the portable movement model described in this section, a simple relationship among λ_h, p_o, p_f, and η was derived in[21]. This idea is described below. Suppose that a portable moves across K cell boundaries during a call holding time assuming that the call is completed. That is, K is the number of hand-offs before the call is completed. The call is referred to as a K-*hand-off call*. If we modify our portable movement model such that t_c represents a call holding time (and λ is replaced by μ in (11.6)), then $\alpha(K)$ is the probability of a K-hand-off call. For a K-hand-off call, let J be the number of portable moves before the call is blocked or successfully terminated, where $J \leq K$. The probability $\alpha(K)$ can be used to derive the expected number $E[J]$. The idea is the following. We first express the conditional probability $\Pr[J = j | K = k]$ as a function of p_f. Then $\Pr[J = j | K = k]$ is used to derive $E[J | K = k]$, the expected number of J for a k-hand-off call.

$$E[J|K=k] = \sum_{j=0}^{k} j \Pr[J=j|K=k] = \frac{1-(1-p_f)^k}{p_f}$$

Finally, $E[J]$ is derived by using $E[J|K=k]$ and $\alpha(k)$:

$$E[J] = \sum_{k=1}^{\infty} E[J|K=k]\alpha(k) = \frac{\eta[1-f^*(\mu)]}{\mu[1-(1-p_f)f^*(\mu)]}$$

where f^* is defined in (11.7). The details of derivations for $\Pr[J = j|K=k]$, $E[J|K=k]$, and $E[J]$ are given in[21]. By assuming homogeneous cells in a PCS network, λ_h is expressed as

$$\lambda_h = (1-p_o)E[J]\lambda_o \qquad (11.9)$$

Thus, we may derive p_o and p_f by an iterative process: Select an initial λ_h value to obtain p_o and p_f for a particular hand-off strategy (analytic models have been developed to derive p_o and p_f for the non-prioritized scheme, the guard channel scheme, and the queueing priority schemes[21]). Then (11.9) is used to compute the new λ_h value which is used to obtain new values for p_o and p_f. The process iterates until the value for λ_h converges.

Figures 11.4 compares p_{nc} (the probability that a call cannot complete due to blocking or forced termination) for the analytic results and the simulation results. The curves indicate that the analytic results are consistent with the simulation experiments.

11.4. CONCLUSIONS

This article described analytic approaches in developing practical models for large-scale PCS networks,

Based on the flow equivalent assumption (the rate of portables move into a cell equals to the rate of portables move out of the cell), a portable population model was described. The model provides the steady-state portable population distribution in a cell which is independent of the portable residence time distribution, which can be used by simulations to reduce the necessary execution time by reaching the steady state more rapidly. Additionally, this model can be used to study the blocking probability of low (portable) mobility PCS network and the performance of portable de-registration strategies.

Figure 11.4: The comparison of p_{nc} for the analytic results and the simulation results (η equals to the call completion rate, and the number of channels in a cell is 10)

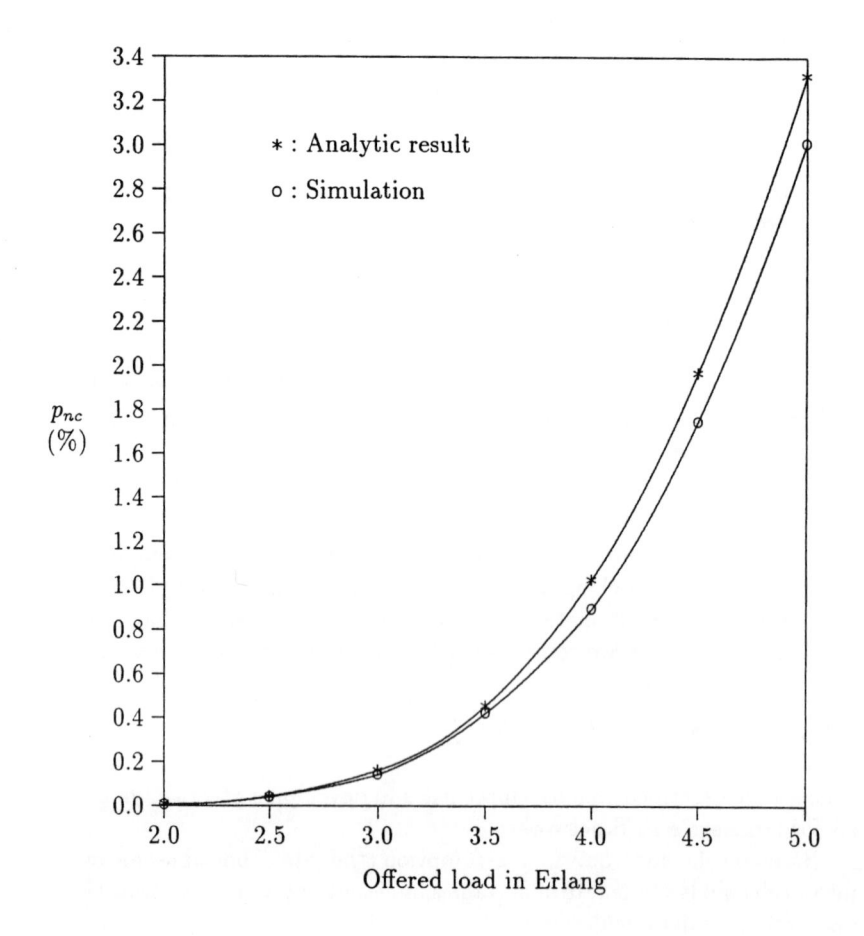

Then we describe a model for portable movement. The model assumes that the arrival calls to a portable form a Poisson process, and portable residence times have a general distribution. This model can be used to study location tracking algorithms and hand-off algorithms. We showed that under some assumptions, the analytic techniques are consistent with the simulation model.

ACKNOWLEDGEMENT

This work is a part of a research project the author collaborated with Christopher D. Carothers and Richard M. Fujimoto.

REFERENCES

References

[1] BELLCORE. "Network and Operations Plan for Access Services to Personal Communications Services Systems, Issue 2. Technical Report SR-TSV-002459, Bellcore (1992).

[2] BELLCORE. "Generic Criteria for Version 0.1 Wireless Access Communications Systems (WACS) and Supplement", Technical Report TR-INS-001313, Issue 1, Bellcore (1994).

[3] D.C. COX, "Wireless Personal Communications: What Is It?" *IEEE Personal Commun. Mag.*, pp. 20-35 (April 1995).

[4] EIA/TIA. "Cellular Radio-Telecommunications Intersystem Operations", Technical Report IS-41 Revision B, EIA/TIA (1991).

[5] S.M. ELNOUBI, R. SINGH, and S.C. GUPTA, "A New Frequency Channel Assignment Algorithm in High Capacity Mobile Communication Systems", *IEEE Trans. Veh. Technol.*, **VT-31**, 3, pp. 125-131 (1982).

[6] G.J. FOSCHINI, B. GOPINATH, and Z. MILJANIC, "Channel Cost of Mobility", *IEEE Trans. Veh. Technol.*, **42**, 4, pp. 414-424 (November 1993).

[7] D. HONG and S.S. RAPPAPORT, "Traffic Model and Performance Analysis for Cellular Mobile Radio Telephone Systems with Prioritized and No-protection Handoff Procedure", *IEEE Trans. Veh. Techol.*, **VT-35**, 3, pp. 77-92 (August 1986).

[8] R. JAIN, Y.-B. LIN, C.N. LO, and S. MOHAN, "A Caching Strategy to Reduce Network Impacts of PCS", *IEEE J. Select. Areas Commun.*, **12**, 8, pp. 1434-1445 (1994).

[9] T.J. KAHWA and N.D. GEORGANAS, "A Hybrid Channel Assignment Scheme in Large-Scale, Cellular-Structure Mobile Communication Systems", *IEEE Trans. Commun.*, **COM-26**, 4, pp. 432-438 (April 1978).

[10] L. KLEINROCK, *Queueing Systems: Volume I - Theory*. New York: Wiley (1976).

[11] S.S. KUEK and W.C. WONG, "Ordered Dynamic Channel Assignment Scheme with Reassignment in Highway Microcells", *IEEE Trans. Veh. Technol.*, **41**, 3, pp. 271-277 (1992).

[12] Y.-B. LIN, "Reducing Location Update Cost in a PCS Network", Submitted for Publication.

[13] Y.-B. LIN, "Determining the User Locations for Personal Communications Networks", *IEEE Trans. Veh. Technol* **43**, 3, pp. 466-473 (1994).

[14] Y.-B. LIN and W. CHEN, "Reducing Call Blocking Probability in a PCS Network Using Call Request Buffering", Technical Report TM-ARH-023161, Bellcore (1994).

[15] Y.-B. LIN and W. CHEN, "A Simulation Model for Hybrid Channel Assignment in a PCS Network", *Intl. J. in Computer Simulation*, **5**, 1, pp. 1-12 (1995).

[16] Y.-B. LIN and S.K. DEVRIES, "PCS Network Signaling Using SS7", *IEEE Personal Commun. Mag.*, pp. 44-55 (June 1995).

[17] Y.-B. LIN and S.-Y. HWANG, "Deregistration Strategies for PCS Networks", Submitted for Publication.

[18] Y.-B. LIN and V.K. MAK, "Eliminating the Boundary Effect of a Large-scale Personal Communication Service Network Simulation", *ACM Tran. on Modeling and Computer Simulation*, **4**, 2 (1994).

[19] Y.-B. LIN and A. NOERPEL, "Implicit Deregistration in a PCS Network", *IEEE Trans. Veh. Technol.*, **43**, 4, pp. 1006-1010 (1994).

[20] Y.-B. LIN, A. NOERPEL, L.F. CHANG, and K.I. PARK, "Performance Modeling of Multi-Tier PCS System", To appear in *Intl. J. of Wireless Information Networks*.

[21] Y.-B. LIN, S. MOHAN, and A. NOERPEL, "Queueing Priority Channel Assignment Strategies for Handoff and Initial Access for a PCS Network", *IEEE Trans. Veh. Technol.*, **43** 3, pp. 704-712 (1994).

[22] E.J. MUTH, *Transform Methods With Applications to Engineering and Operations Research*. Prentice-Hall (1977).

[23] P. PORTER, D. HARASTY, M. BELLER, A. NOERPEL, and V. VARMA, "The Terminal Registration/Deregistration Protocol for Personal Communication Systems", *Wireless 93 Conf. on Wireless Commun.* (July 1993).

[24] S.M. ROSS, *Stochastic Processes*. John Wiley & Sons (1983).

[25] S. TEKINARY and B. JABBARI, "Handover Policies and Channel Assignment Strategies in Mobile Cellular Networks", *IEEE Commun. Mag.*, **29**, 11 (1991).

[26] S. TEKINARY and B. JABBARI, "A Measurement Based Prioritization Scheme for Handovers in Cellular and Microcellular Networks", *IEEE J. Select. Areas Commun.*, pp. 1343-1350 (October 1992).

[27] E.J. WATSON, *Laplace Transforms and Applications*. Birkhauserk 1981.

[28] C.H. YOON and K. UN, Performance of personal portable radio telephone systems with and without guard channels. *IEEE J. Select. Areas Commun.*, **11**, 6, pp. 911-917 (August 1993).

[29] M. ZHANG and T.-S. YUM, "Comparisons of Channel-Assignment Strategies in Cellular Mobile Telephone Systems", *IEEE Trans. Veh. Technol.*, **38**, 4, pp. 211-215 (1989).

CHAPTER 12

THE EFFECT OF REQUEST WAITING IN A MOBILE COMPUTING NETWORK

Yi-Bing Lin and Wai Chen

12.1. INTRODUCTION

In a *mobile computing network* (MCN), the service area is populated with *mobile data base stations* (MDBS), each providing coverage in its vicinity. Each MDBS is assigned a group of channels. When a subscriber or *mobile end station* (M-ES) connects to the MCN, it consumes a channel until the end of the computing session. If no channel is available, the access request is dropped. In many cases, channels may return shortly after an access request is dropped. Thus, if some buffering mechanism is introduced to the channel allocation algorithm, a cell may accommodate more requests. This is referred to as *request waiting* or *request buffering*. A timeout constraint may or may not be associated with the buffering mechanism. Theoretically, a waiting request is eventually connected under the FIFO policy if there is no timeout constraint. In practice, a waiting M-ES may be impatient and decide to drop the request after a waiting period.

A mechanism similar to request waiting is *user re-dialing*. When an M-ES fails to access a channel, the user may re-try (re-dial) after a short period of time. The advantage of the user re-dial is to eliminate the buffering mechanism maintained by the MCN. On the other hand, user re-dialing requires an M-ES to repeatedly initiate the request process.

This chapter studies the performance of the request buffering, and compares it with the user re-dial. The chapter is organized as follows. Section 12. proposes an M-ES mobility model. Based on the mobility model, Section 12. analyzes the request buffering mechanism. Section 12. studies user re-dialing mechanism and compares request waiting with user re-dialing. The results indicate that the performance of request waiting is much better than user re-dialing.

12.2. THE MOBILITY MODEL

Our mobility model for an MCN is different from the models for the existing proposed *personal communications services* (PCS) networks[5, 8, 7]. In a PCS network, it is likely that a *portable* (a PCS subscriber) crosses several cell boundaries during a phone conversation, and the *hand-off* behavior must be addressed in the mobility model.

In an MCN, wireless channels cannot offer the stability of their wireline equivalents (because of signal fading, noise, and other difficulties). Due to the M-ES movement, the channel bit rate can drop dramatically. The usual solution is for a user to find a spot with acceptable transmission quality and move after the computation session is complete[2] Our model assume that no computation session is in progress when the M-ES is crossing the cell boundaries, and hand-offs in the PCS models do not exist. The reader is referred to[5] if hand-offs should be considered. In this chapter, mobility is addressed by the user population distribution at a cell as proposed in Chapter 11 of this book.

Let N be the expected number of users in a cell. Suppose that the resident time of a user in a cell (i.e., the period that a user stays in a cell) has a general distribution $F(t)$ with mean $\frac{1}{\eta}$. In the steady state, the rate of the users move in a cell equals to the rate of the users move out of the cell. In other words, the rate that users move in a cell is $N\eta$. The arrivals of users can be viewed as being generated from N input streams which have the same general distribution with arrival rate η. If N is reasonably large ($N \geq 40$) in an MCN, the net input stream is approximated as a Poisson process with arrival rate $N\eta$. Thus, the distribution for the user population can be modeled by an $M/G/\infty$ queue with arrival rate $N\eta$ and the leaving rate (or service rate in the queueing theory terminology)η. Let π_n be the steady state probability that there are n users in the cell. From the standard technique[3],

$$\pi_n = \frac{N^n e^{-N}}{n!} \qquad (12.1)$$

The user distribution (i.e., Equation (12.1)) is used to study the blocking probability p_b (the probability that an access request is dropped) for an MCN. The idea was described in Chapter 11, and is re-iterated here. We derive the blocking probability $p_b^{(n)}$ for a cell when there are n users. By using (12.1), the expected blocking probability for the cell is

$$p_b = \sum_{c < n < \infty} \pi_n p_b^{(n)} \qquad (12.2)$$

where c is the number of channels in the cell.

12.3. PERFORMANCE MODELING FOR REQUEST WAITING

We propose both simulation and analytic models for request waiting. This section ddescribes the assumptions and the results of our request waiting model. The details of the simulation and analysis are given in Appendices 12.I, 12.II, and 12.III.

We consider a buffering mechanism with fixed and exponential timeouts. The exponential timeout can be used to model the situation where the buffering mechanism does not have a timeout limit, but the users may be impatient, and decide to drop the request after an exponential waiting time.

Our model makes the following assumptions:

Assumption 1. The request inter arrival time for an M·ES is exponentially distributed with mean $1/\lambda$.

Assumption 2. The computing session time or the channel occupancy time is exponentially distributed with mean $1/\mu$.

Assumption 3. The timeout period is a constant τ in the fixed timeout model. In the exponential timeout model, the timeout period is exponentially distributed with mean $\tau = 1/\gamma$.

Assumption 4. The waiting requests are queued in the buffering mechanism in the first-come-first-serve order.

Note that in our simulation model, distributions other than exponential can be used in Assumptions 1-3. The exponential distributions (and the fixed timeout period) are selected for demonstration and validation purposes.

Figure 12.1 compares the fixed timeout model with the exponential timeout model. We observe the following.

- The blocking probability p_b decreases as the mean timeout period τ increases.

- There exists a timeout value τ^* such that for $\tau < \tau^*$, the performance of the exponential timeout is better than the fixed timeout. We note that τ^* increases as λ increases ($\tau^* \simeq 0.45/\mu$ for $\lambda = 0.15\mu$, $\tau^* \simeq 0.28/\mu$ for $\lambda = 0.125\mu$, and $\tau^* \simeq 0.1/\mu$ for $\lambda = 0.1\mu$). We also note that if $\tau^* > 10/\mu$, p_b is insensitive to the timeout distributions[5].

Figure 12.2 shows the effect of the mean timeout period (the exponential timeout model is considered). As mentioned before, the blocking probability p_b decreases as τ increases. In fact, $\lim_{\tau \to \infty} p_b = 0$. In this figure, the workload of the mobile computing network with $\tau = 0$ is engineered at 1% blocking probability. The workload increase in the figure is with respect to the engineered workload for $\tau = 0$. Figure 12.2 (a) indicates that the MCN can carry 20%, 30%, 40%, and 60% more offered load if the timeout period is increased from 0 to $0.1/\mu, 0.2/\mu, 0.3/\mu$, and $0.5/\mu$ respectively (the blocking probability is maintained at 1%). Figure 12.2 (b) shows the expected waiting time $E[\tau_1]$ of an M-SE if it is eventually connected. The figure indicates that if an M-SE is connected, it is expected to be connected in a short period of time.

Based on Figure 12.2, one may tend to draw the conclusion that we should select a large timeout period because it improves network performance within a reasonable connection waiting time. However, in many cases, the service quality must be guaranteed by limiting the worst case waiting time. In other words, the waiting time of the rejected users must be considered. To provide low blocking probability and short worst case waiting time, it is necessary to increase the number of channels. Figure 12.3 compares the channel cost and the waiting cost when the workload increases. In this figure, the workload is engineered at 1% blocking probability for $\tau = 0.1/\mu$ where the number of channels for a cell is $c = 8$. When the workload of the network increases by

REQUEST WAITING IN MOBILE COMPUTING

Figure 12.1: The comparison of the fixed timeout and the exponential timeout

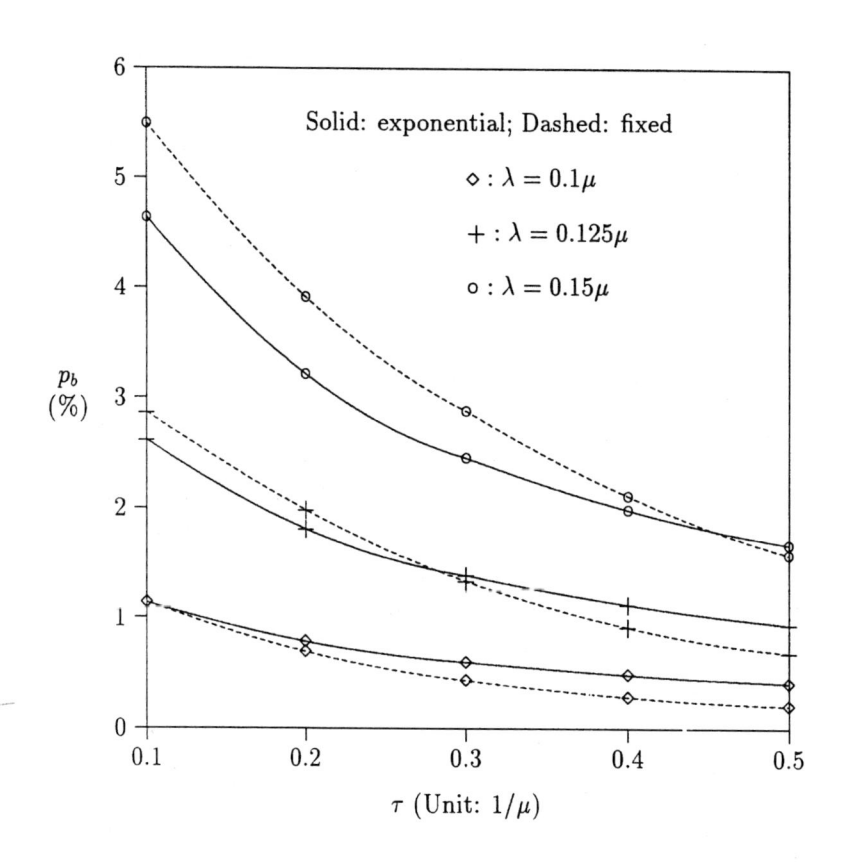

Figure 12.2: The effect of timeout (the exponential model)

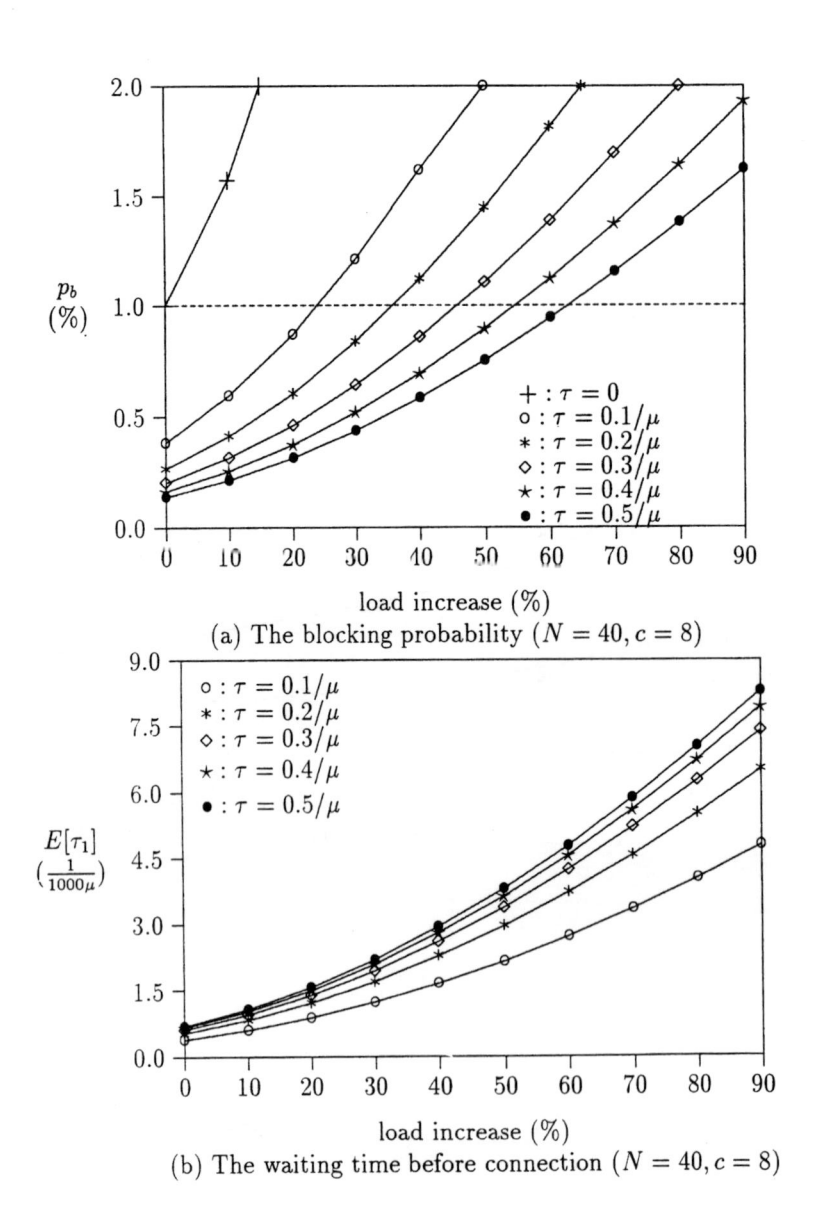

p_b (%)

$+ : \tau = 0$
$\circ : \tau = 0.1/\mu$
$* : \tau = 0.2/\mu$
$\diamond : \tau = 0.3/\mu$
$\star : \tau = 0.4/\mu$
$\bullet : \tau = 0.5/\mu$

load increase (%)
(a) The blocking probability ($N = 40, c = 8$)

$\circ : \tau = 0.1/\mu$
$* : \tau = 0.2/\mu$
$\diamond : \tau = 0.3/\mu$
$\star : \tau = 0.4/\mu$
$\bullet : \tau = 0.5/\mu$

$E[\tau_1]$ ($\frac{1}{1000\mu}$)

load increase (%)
(b) The waiting time before connection ($N = 40, c = 8$)

Figure 12.3: Channel cost vs. waiting cost

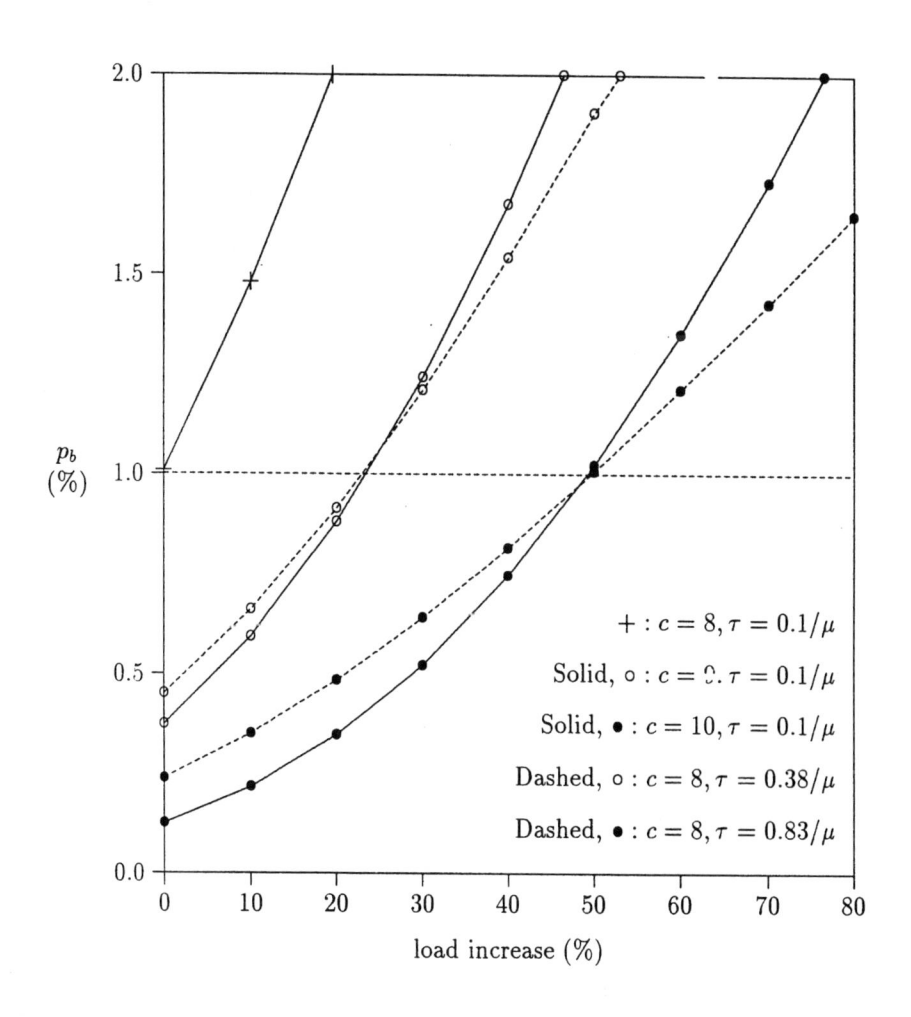

p_b (%)

load increase (%)

$+ : c = 8, \tau = 0.1/\mu$

Solid, $\circ : c = 8, \tau = 0.1/\mu$

Solid, $\bullet : c = 10, \tau = 0.1/\mu$

Dashed, $\circ : c = 8, \tau = 0.38/\mu$

Dashed, $\bullet : c = 8, \tau = 0.83/\mu$

20%, the same blocking probability can be maintained if the channel number is increased from 8 to 9 or if the timeout period is increased from $0.1/\mu$ to $0.38/\mu$. Similarly, if the workload increases by 50%, the same blocking probability can be maintained by using 10 channels or increasing timeout period to $0.83/\mu$.

12.4. THE EFFECT OF USER RE-DIALING

This section describes the effect of user re-dialing. In the user re-dialing model, we use Assumptions 1 and 2 as well as the following assumption.

Assumption 5. If an M-SE fails to access a channel, it may decide not to re-try (with probability $1 - \beta$) or it may decide to re-try (with probability β) after a period which is exponentially distributed with mean $1/\lambda_r$.

The simulation model for user re-dialing is constructed by modifying the simulation model described in Appendix 12.III. The details of the simulation are omitted.

Figure 12.4 illustrates the impact of user re-dialing. Figure 12.4 (a) indicates that re-dialing does improve the performance, and p_b decreases as β increases or as λ_r decreases. Figure 12.4 (b) indicates that the performance improvement by re-dialing become less significant as λ increases.

To compare the performance of request waiting and user re-dialing, we first compute the expected time τ_r before a user decides not to re-try. With the same call arrival rate, we compare the blocking probability for request waiting (with timeout period τ) with that of user re-dialing (with $\tau_r = \tau$). The time τ_r is derived as follows.

$$\begin{aligned}\tau_r &= \frac{1}{\lambda_r} \sum_{i=1}^{\infty} i(1-\beta)\beta^{i-1} \\ &= \frac{1}{\lambda_r(1-\beta)}\end{aligned} \quad (12.3)$$

Consider Figure 12.4 (a). The lowest blocking probability $p_b = 2.6\%$ when $\lambda_r = 5\lambda = 0.625\mu$ and $\beta = 0.5$. In other words, $p_b = 2.6\%$ when $\tau_r = 3.2/\mu$. Consider the curves for $\lambda = 0.125\mu$ in Figure 12.1. We observe that $p_b = 0.94\%$ for $\tau = 0.5/\mu$ (the exponential timeout model) and $p_b = 0.68\%$ for $\tau = 0.5/\mu$ (the fixed timeout model). In

REQUEST WAITING IN MOBILE COMPUTING 249

Figure 12.4: The Effect of user re-dialing ($N = 40, c = 8$)

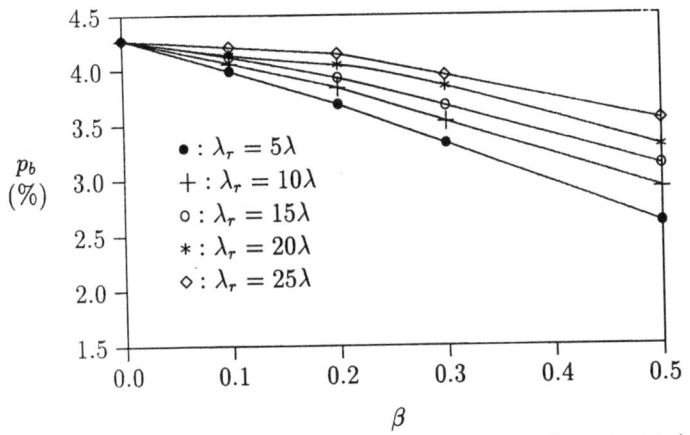

(a) The effect of the re-dial probability ($\lambda = 0.125\mu$)

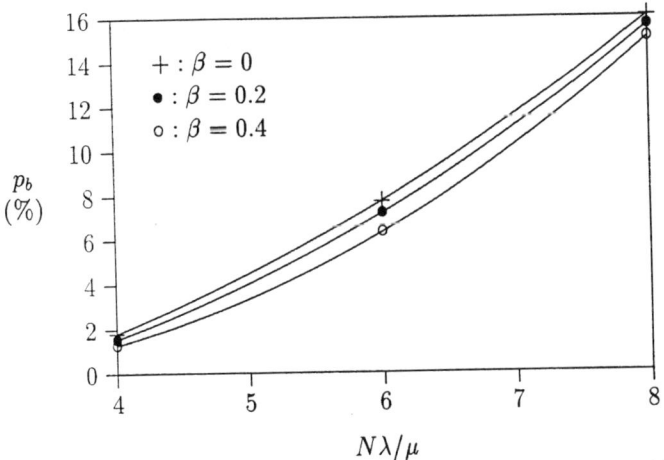

- $\bullet : \lambda_r = 5\lambda$
- $+ : \lambda_r = 10\lambda$
- $\circ : \lambda_r = 15\lambda$
- $* : \lambda_r = 20\lambda$
- $\diamond : \lambda_r = 25\lambda$

(b) The effect of the request arrival rate ($\lambda_r = 10\lambda$)

- $+ : \beta = 0$
- $\bullet : \beta = 0.2$
- $\circ : \beta = 0.4$

other words, when $\tau = 3.2/\mu$, $p_b < 0.94\%$ for request waiting, and we conclude that with the same expected waiting/re-dialing period (τ or τ_r), the performance for request waiting is much better than user re-dialing.

12.5. SUMMERY

This chapter proposed a buffering mechanism for a mobile computing network (the approach is called *request waiting*). If no idle channel is available when an access request arrives, the request is buffered in a waiting queue. The waiting request is either accepted if a channel is available or rejected after a timeout period (either defined by the network or by the user). This mechanism eliminates the need of user re-dialing.

Our results indicate that the performance of request waiting is much better than user re-dialing with the same timeout period. We also showed how to limit the user waiting time by controlling the number of channels at a cell. One of the future research directions is to consider the possible variations in channel bit rate[6].

References

[1] Gnedenko, B.V. and Kovalenko, I.N. *Introduction to Queueing Theory*. Program for Scientific Translation, Jerusalem, Israel, 1968.

[2] K. Miller. Cellular Essentials for Wireless Data Transmission. *Data Communications*, March 1994.

[3] Kleinrock, L. *Queueing Systems: Volume I - Theory*. New York: Wiley, 1976.

[4] Lin, Y.-B. and Chen, W. Impact of busy lines and mobility on cell blocking in a PCS network. *ICCCN '94*, pages 343–347, 1994.

[5] Lin, Y.-B., and Chen, W. Reducing Call Blocking Probability in a PCS Network Using Call Request Buffering. Technical Report TM-ARH-023161, Bellcore, 1994. To appear in *IEEE Personal Communications Magazine*.

[6] Lin, Y.-B., Lin, Y.-J., and Mak, V.W.K. Allocating Resources for Soft Requests – A Performance Study. *Information Sciences*, 85(1):39–65, 1995.

[7] Lin, Y.-B., Mohan, S., and Noerpel, A. Channel Assignment Strategies for Hand-off and Initial Access for a PCS Network. *IEEE Personal Communications Magazine*, 1(3):47–56, 1994.

[8] Lin, Y.-B., Mohan, S., and Noerpel, A. Queueing Priority Channel Assignment Strategies for Handoff and Initial Access for a PCS Network. *IEEE Trans. Veh. Technol.*, 43(3):704–712, 1994.

APPENDIX 12.I. THE FIXED TIMEOUT MODEL

This section proposes an analytical model to study the performance for request waiting with fixed timeout. We assume low c/N value.

12.I.1. The Blocking Probability

Suppose that a call C_t arrives at time t. After the call arrives, let X_i be the time period that channel i is busy before all calls preceding C_t have left the system (either completed or dropped). Let $p(x_1, x_2, ..., x_c)$ be the density function that $X_i = x_i$ when C_t arrives. If the number of portables in a cell is $n \gg c$, then we may assume fixed call stream $n\lambda$ to the cell. From the standard technique[1],

$$p(x_1, x_2, ..., x_c) = \alpha(n) \frac{(n\lambda)^c}{c!} e^{-\mu(x_1+...+x_c)+n\lambda \min(\tau, x_1,...,x_c)}$$

where

$$\alpha(n) = \sum_{0 \le k \le c-1} \frac{(n\lambda)^k}{k! \mu^k}$$

$$+ \begin{cases} \dfrac{(n\lambda)^c}{c! \mu^c} \left[\dfrac{n\lambda e^{-\tau(c\mu-n\lambda)} - c\mu}{n\lambda - c\mu} \right] & \text{for } n\lambda \ne c\mu \\ \dfrac{c^c}{c!}(1 + n\lambda\tau) & \text{for } n\lambda = c\mu \end{cases}$$

A phone call is blocked if $x_i > \tau$ for $1 \leq i \leq c$. Thus, the blocking probability $p_b^{(n)}$ is

$$\begin{aligned}
p_b^{(n)} &= \Pr[\tau < \min(X_1, ..., X_c)] \\
&= \int_{x_1=\tau}^{\infty} \cdots \int_{x_c=\tau}^{\infty} p(x_1, ..., x_c) dx_1 ... dx_c \\
&= \int_{x_1=\tau}^{\infty} \cdots \int_{x_c=\tau}^{\infty} \frac{\alpha(n) n^c \lambda^c}{c!} e^{-\mu(x_1+...+x_c)+n\lambda\tau} dx_1 ... dx_c \\
&= \frac{\alpha(n) n^c \lambda^c}{c! \mu^c} e^{-\tau(c\mu - n\lambda)}
\end{aligned}$$

From (12.2), the blocking probability p_b for low mobility portables is

$$\begin{aligned}
p_b &= \sum_{c<n<\infty} \pi_n p_b^{(n)} \\
&= \frac{\lambda^c}{c!\mu^c} e^{-\tau c\mu} \left[\sum_{c<n<\infty} \frac{\alpha(n) n^c N^n}{n!} e^{-(N-n\tau\lambda)} \right] \quad (12.4)
\end{aligned}$$

Figure 12.5 (a) compares Equation (12.4) against the simulation experiments in Appendix 12.III for different τ values. In the experiments, the average number of users is $N = 200$, the number of channels is $c = 5$, and the traffic per user is $\lambda/\mu = 0.01$ Erlang. The figure indicates that the analytical analysis is in agreement with the simulation study.

12.I.2. The Expected Waiting Time

Suppose that there are n portables in a cell. Let τ' be the waiting time for a call ($0 \leq \tau' < \tau$ if the call is connected; otherwise the call is dropped). The density function $f_n(\tau')$ for $0 < \tau' < \tau$ is

$$\begin{aligned}
f_n(\tau') &= \binom{c}{1} \int_{x_2=\tau'}^{\infty} \cdots \int_{x_c=\tau'}^{\infty} p(\tau', x_2, ..., x_c) dx_2 ... dx_c \\
&= \binom{c}{1} \int_{x_2=\tau'}^{\infty} \cdots \int_{x_c=\tau'}^{\infty} \alpha(n) \frac{(n\lambda)^c}{c!} e^{-\mu(\tau'+x_2+...+x_c)+n\lambda\tau'} dx_2 ... dx_c \\
&= \frac{c(n\lambda)^c \alpha(n)}{\mu^{c-1} c!} e^{-(c\mu - n\lambda)\tau'}
\end{aligned}$$

The probability $\Pr[\tau' = \tau] = p_b^{(n)}$ is given in (12.4). Suppose that a phone call is connected. Then the mean waiting time $E_n[\tau_1]$ until the

REQUEST WAITING IN MOBILE COMPUTING

Figure 12.5: The comparison of the analytic model and the simulation model for the fixed timeout model ($c = 5, N = 200, \lambda/\mu = 0.01$)

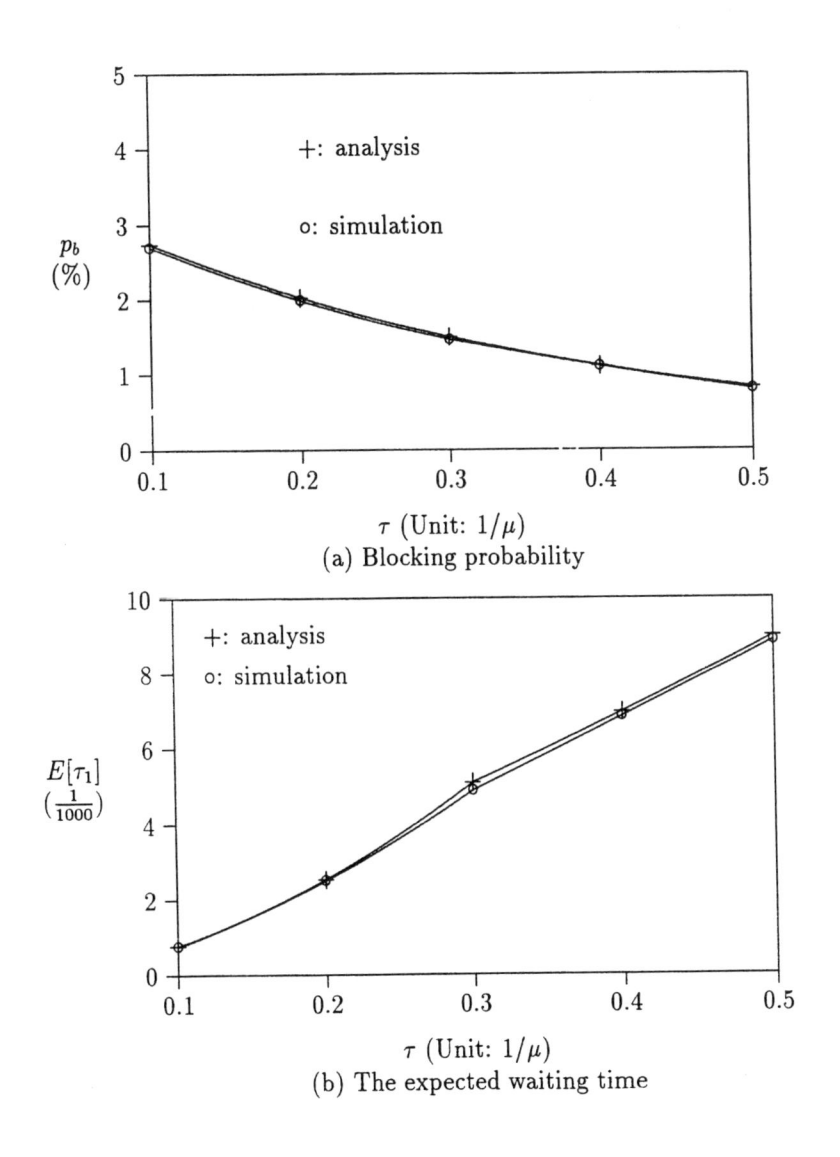

(a) Blocking probability

(b) The expected waiting time

beginning of the connection is

$$
\begin{aligned}
E_n[\tau_1] &= E_n[\tau'|\tau' < \tau] \\
&= \left\{\frac{1}{1 - \Pr[\tau' = \tau]}\right\}\left\{0 \times \Pr[\tau' = 0] + \int_{\tau'=0+}^{\tau} \tau' f_n(\tau')d\tau'\right\} \\
&= \begin{cases} \dfrac{c\mu(n\lambda)^c \alpha(n)\left\{1 - e^{-(c\mu-n\lambda)\tau}[c\mu + n\lambda\tau(c\mu - n\lambda)]\right\}}{\left[c!\mu^c - \alpha(n)n^c\lambda^c e^{-\tau(c\mu-n\lambda)}\right](c\mu - n\lambda)^2} & \text{for } c\mu = n\lambda \\[1em] \dfrac{c\mu(n\lambda)^c \alpha(n)\tau^2}{2\left[c!\mu^c - \alpha(n)n^c\lambda^c\right]} & \text{for } c\mu \neq n\lambda \end{cases}
\end{aligned}
\quad (12.5)
$$

For a call being connected, the expected waiting time $E[\tau_1]$ in the low-mobility model is

$$
\begin{aligned}
E[\tau_1] &= \sum_{c < n < \infty} \frac{E_n[\tau'|\tau' < \tau] N^n e^{-N}}{n!} \\
&= \begin{cases} \displaystyle\sum_{n=c+1}^{\infty} \frac{c\mu(n\lambda)^c \alpha(n)\left\{1 - e^{-(c\mu-n\lambda)\tau}[c\mu + n\lambda\tau(c\mu - n\lambda)]\right\} N^n e^{-N}}{\left[c!\mu^c - \alpha(n)(n\lambda)^c e^{-\tau(c\mu-n\lambda)}\right](c\mu - n\lambda)^2 n!}, \\ \hfill \text{for } c\mu = n\lambda \\[1em] \displaystyle\sum_{n=c+1}^{\infty} \frac{c\mu(n\lambda)^c \alpha(n)\tau^2 N^n e^{-N}}{2\left[c!\mu^c - \alpha(n)(n\lambda)^c\right]n!}, \\ \hfill \text{for } c\mu \neq n\lambda \end{cases}
\end{aligned}
\quad (12.6)
$$

Figure 12.5 (b) compares (12.6) with the simulation experiments for different τ values. The figure indicates that the analytic model is consistent with the simulation model.

APPENDIX 12.II. THE EXPONENTIAL TIMEOUT DISTRIBUTION MODEL

The exponential timeout model for request waiting is similar to the one developed in[8]. The reader is referred to[8] and[5] for the detailed derivations. We only show the results for completeness.

REQUEST WAITING IN MOBILE COMPUTING   255

If the timeout period has an exponential distribution with mean $\tau = 1/\gamma$, the blocking probability is (see Equation (7) in[5])

$$p_b = \sum_{c < n < \infty} \left[\sum_{0 \leq j < n-c} \frac{(j+1)\gamma p_{c,j}^{(n)}}{c\mu + (j+1)\gamma} \right] \frac{N^n e^{-N}}{n!} \qquad (12.7)$$

where

$$p_{i,j}^{(n)} = \begin{cases} \left\{ 1 + \sum_{i=1}^{c} \frac{n!}{(n-i)!i!} \left(\frac{\lambda}{\mu}\right)^i + \left[\frac{n!}{(n-c)!c!} \left(\frac{\lambda}{\mu}\right)^c\right] \sum_{j=1}^{n-c} \frac{(n-c)! \lambda^j}{j! \prod_{1 \leq m \leq j}(c\mu + m\gamma)} \right\}^{-1}, \\ \qquad\qquad\qquad\qquad\qquad\qquad\qquad\qquad \text{for } i = j = 0 \\[2pt] \frac{n!}{(n-i)!i!} \left(\frac{\lambda}{\mu}\right)^i p_{0,0}^{(n)}, \\ \qquad\qquad\qquad\qquad\qquad\qquad\qquad\qquad \text{for } 1 \leq i \leq c, j = 0 \\[2pt] \frac{(n-c)!\lambda^j}{j! \prod_{1 \leq m \leq j}(c\mu + m\gamma)} p_{c,0}^{(n)}, \\ \qquad\qquad\qquad\qquad\qquad\qquad\qquad\qquad \text{for } i = c, 1 \leq j \leq n-c \end{cases}$$

Figure 12.6 (a) compares Equation (12.7) with the simulation experiments. In the figure, the average user number is $N = 40$, the number of channels is $c = 8$, the traffic per user is $\lambda/\mu = 0.1$ Erlang. The analytical results are consistent with the simulation experiments.

The expected waiting time $E[\tau_1]$ for an accepted request in the exponential timeout model is (see Equation (8) in[5])

$$E[\tau_1] = \sum_{c<n<\infty} \frac{\left(\sum_{0 \leq j < n-c} \sum_{0 \leq i \leq j+1} \frac{p_{c,j}^{(n)}}{c\mu + i\gamma} \right) N^n e^{-N}}{\left[1 - \sum_{0 \leq j < n-c} \frac{(j+1)\gamma p_{c,j}^{(n)}}{c\mu + (j+1)\gamma} \right] n!} \qquad (12.8)$$

Figure 12.6 (b) plots $E[\tau_1]$ based on Equation (12.8) (the curves marked •). The analytical analysis is consistent with the simulation study (the curves marked o).

Figure 12.6: The comparison of the analytic model and the simulation model for the exponential timeout model ($c = 8, N = 40, \lambda/\mu = 0.1$)

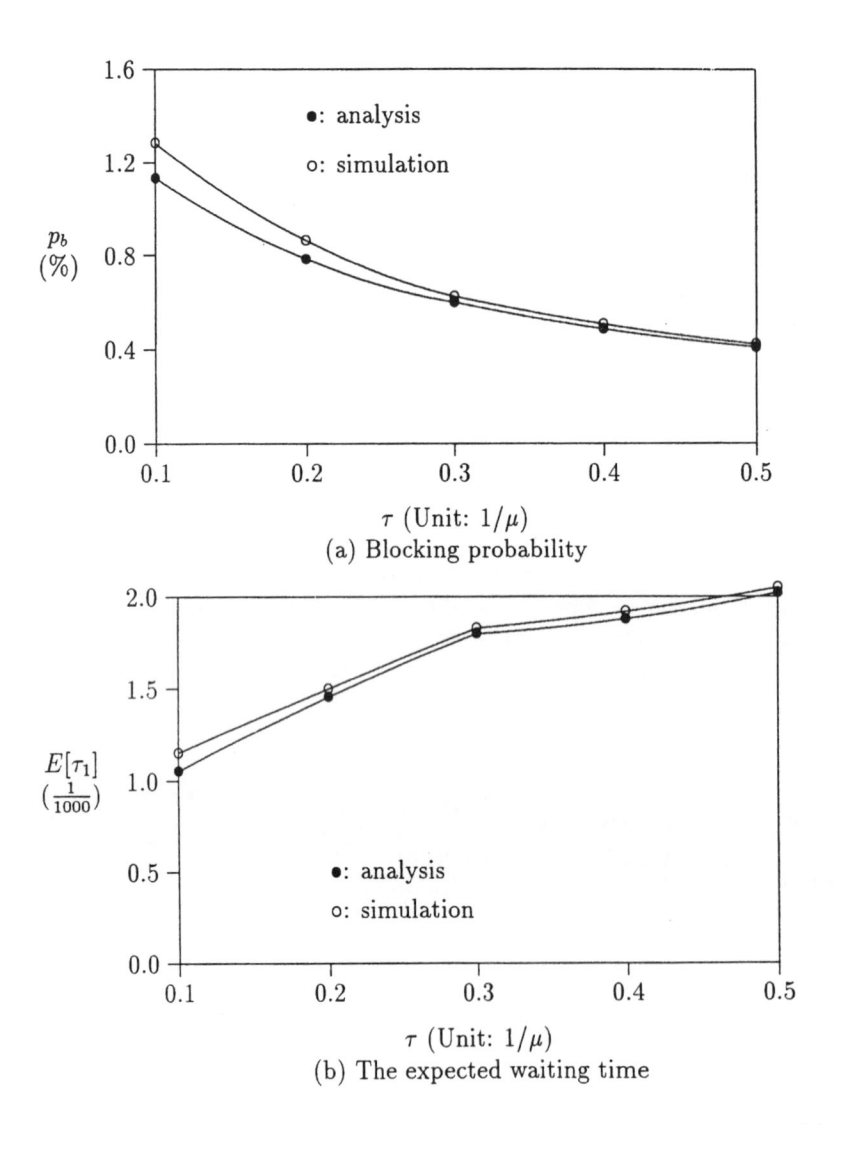

(a) Blocking probability

(b) The expected waiting time

APPENDIX 12.III. THE SIMULATION MODEL FOR REQUEST WAITING

This section describes the simulation model for request waiting. We consider a buffering mechanism with fixed and exponential timeouts. The exponential timeout can be used to model the situation where the buffering mechanism does not have a timeout limit, but the users may be impatient, and decide to drop the request after an exponential waiting time.

Based on Assumptions 1-3, the simulation model works as follows. There are three types of events in the simulation: the ARRIVAL events for request arrivals, the COMPLETION events for users releasing the channels (session completes), and the WAIT events for requests waiting for service. The events are inserted into an event list, and are deleted (and processed) from the event list in the non-decreasing timestamp order. A simulation clock is maintained to indicate the progress of the simulation. The clock value is the timestamp of the event being processed. The simulation iterates $5N - c - 1$ times to compute the blocking probability $p_b^{(n)}$ when there are n users in a cell ($c+1 \leq n \leq 5N$), where c is the number of channels in a cell and N is the expected number of users in a cell. We observe that the probability that more than $5N$ users in a cell is negligible. In an iteration, the simulation counts K, the total number of events being processed. The simulation iteration terminates when $K = 1,000,000$. We observe that at the end of an iteration, the confidence intervals of $p_b^{(n)}$ for the 98% confidence level are within 5% of the mean values. The details of the simulation are given in Figure 12.7. Consider a simulation iteration (Steps 3-20 in Figure 12.7) when n users are in the cell. Step 3 initializes the simulation by generating n ARRIVAL events (one for each user). The event generation is completed in the following steps:

1. Allocate the storage for the event.

2. Determine the type (either ARRIVAL, COMPLETION, or WAIT) of the event.

3. Determine the timestamp of the event. If the event type is ARRIVAL, then the timestamp is the current clock value plus the inter-arrival time (drawn from an exponential distribution with

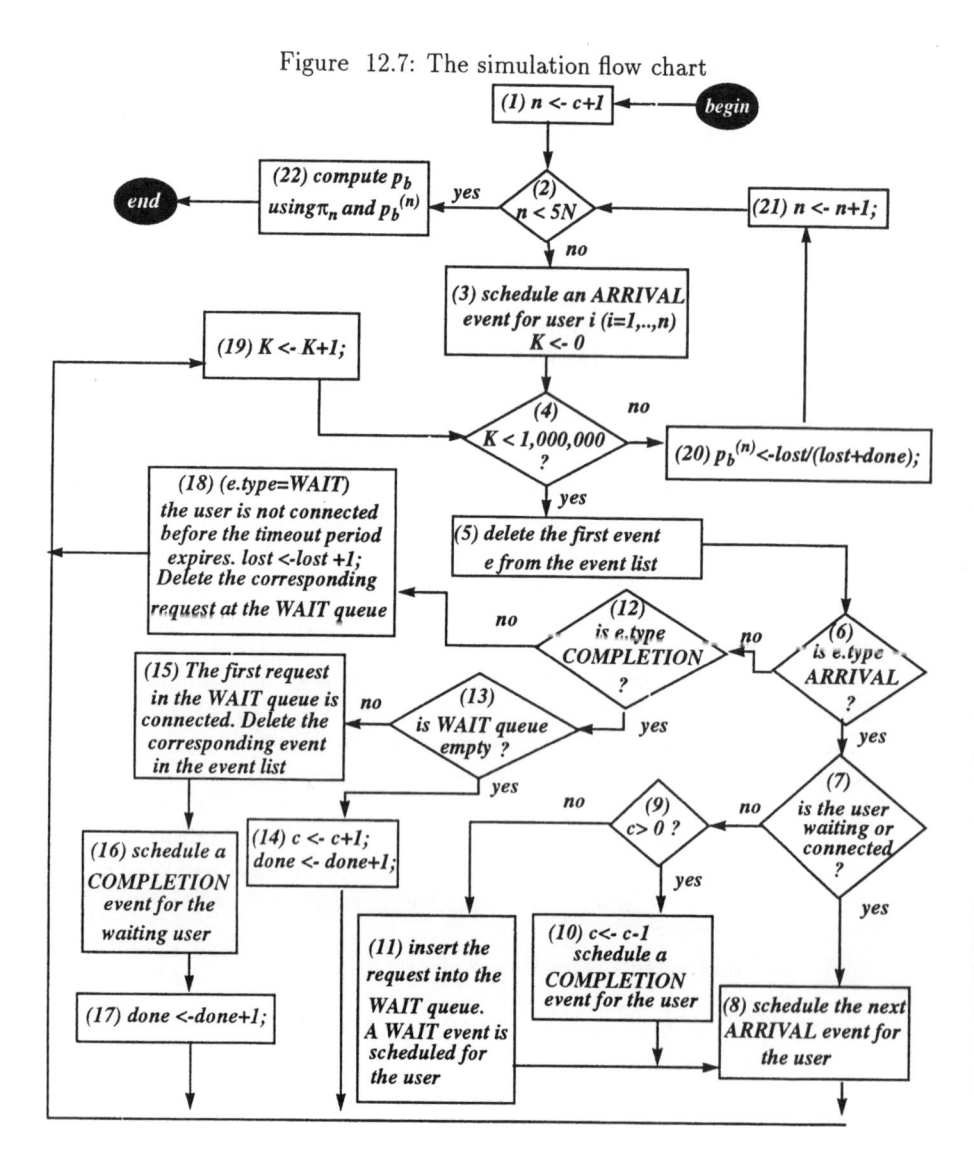

Figure 12.7: The simulation flow chart

mean $1/\lambda$). If the event type is COMPLETION, then the timestamp is the clock plus the channel occupancy time (drawn from an exponential distribution with mean $1/\mu$). If the event type is WAIT, then the timestamp is the clock plus the timeout period (drawn from a distribution with mean $1/\gamma$).

4. Insert the event into the event list.

The next event is deleted from the event list (Step 5), and is processed based on its type (see Steps (6), (12), and (18)).

ARRIVAL (Step 6). Step 7 checks if the user is waiting or is connected for a session. If so, the arrival is ignored. (A user cannot initiates a new session request while he/she is waiting or connected. This phenomenon is referred to as the *busy line* effect[4].) Otherwise, Step 9 checks if a channel is available for the connection. If so, the idle channel is assigned to the user, and a COMPLETION event for the user is generated and inserted in the event list. If no channel is available (Step 11), the request is buffered in the WAIT queue based on a queueing policy (in our experiments, first-come-first-out policy is assumed), and a WAIT event is generated and inserted into the event list. At the end of the ARRIVAL event execution, a new ARRIVAL event for the user is scheduled at Step 8.

COMPLETION (Step 12). A channel is released when a session is completed. If the WAIT queue is empty at Step 13, then the released channel becomes idle (Step 14). Otherwise, a waiting request in the WAIT queue is selected for connection. The corresponding WAIT event in the event list is deleted (because the user is no longer waiting), and a COMPLETION event is scheduled for the waiting user at Step 16. At the end of the COMPLETION event execution, the variable *done* is incremented by one (see Steps 14 and 17).

WAIT (Step 18). The user is not able to obtain a channel before the timeout period expires and the request fails (the *lost* variable is incremented by one). The corresponding request is deleted from the WAIT queue.

After an event is processed, the variable K is incremented by one (Step 19). Step 4 checks if $K = 1000000$. If not, the next event is processed at

Step 5. Otherwise, the current simulation iteration terminates, and $p_b^{(n)}$ is computed at Step 20. The control proceeds to the next simulation iteration for the user number $n+1$ (Step 21). If $n > 5N$ at Step 2 then the simulation completes and the final p_b value is computed at Step 22.

Notes on Contributors

RASSUL AYANI received the Dipl. Ing. degree from the University of Technology in Vienna, Austria, Master's degree from the University of Stockholm and PhD degree from the Royal Institute of Technology (KTH) in Stockholm, Sweden. He is an **Associate Professor** in the Department of Teleinformatics, Royal Institute of Technology. His current research interests are in parallel architectures, parallel algorithms and parallel simulation. He is an Associate Editor of the ACM Transactions on Modeling and Computer Simulation (TOMACS) and is a member of the Editorial Board of the International Journal on Computer Simulation. He is a member of the SCS, IEEE and ACM. His Email address is: rassul@it.kth.se.

KALLOL BAGCHI received his MSc in Mathematics from Calcutta University and Ph.D. in Computer Science in 1988 from Jadavpur University, Calcutta, India. He has worked in industry in India and in Finland. He taught a course at the University of Oulu, Finland in 1986. He worked as an **Assistant** and **Associate Professor** in Computer Science and Engineering at Aalborg University, Denmark from 1987-1992. In 1993, he visited Stanford University, CA. He also completed a certificate in Computer Networking from Columbia University, NY. His interests are in performance modeling and simulation, parallel systems. He has authored or co-authored over 40 international papers in these areas. He has been an associate editor or member of the editorial board of the International Journal in Computer Simulation for the last few years and have guest-edited several issues of the journal. He was a member of the board of directors of SCS in 1993. He has been cited in World Who's who and in Who's who in Science and Engineering. He is a member of the ACM and IEEECS. He has been associated with the MASCOTS workshop, since its inception. At present, he is **pursuing a second Ph.D. degree** in business (DIS) at Florida Atlantic University.

FRANK BALL was awarded a BSc from the Univesity of Lancaster in 1988. He has been employed as a **Research Associate** at Lancaster for the past seven years during which time he has worked on a number of projects relating to quality of service for continuous media. The main focus of his work has been the provision of network level support for continuous media traffic. His main research interests are modelling and performance evaluation of communications networks, and the development of traffic control mechanisms which support guaranteed network services to continuous media.

MICHAEL S. BORELLA was born on Long Island, New York. He received the B.S. degree in Computer Science and Technical Communication (with distinction) from Clarkson University, Potsdam, N.Y., in 1991 and the M.S. and Ph.D. degrees from the University of California, Davis, both in computer science, in 1994 and 1995, respectively. He worked as a technical writer for IBM, Poughkeepsie in 1990, and as a co-op engineer for Rolm, Santa Clara in 1992. From 1992-1995 he was a member of the UC Davis Networks Research Lab, where his research was supported by ARPA and the NSF. He is currently an **Assistant Professor** in the Computer Science Department of DePaul University, in Chicago, Illinois. His research interests are computer security, performance evaluation and modeling of computer networks, architectures and protocols for high-speed wavelength-division-multiplexed local and wide area networks (including multicasting and scheduling issues), and the routing of multicast traffic in regular and random graphs.

GEORG CARLE is currently **finishing his PhD degree** in Computer Science at the University of Karlsruhe, Institute of Telematics, where he participated at the fellowship program 'Controllability of Complex Systems' Phone of the Faculty of Computer Science. He obtained a degree in Electrical Engineering from the University of Stuttgart in 1992, where he performed his diploma project at the Institute for Communication Switching and Data Techniques. In 1990, he performed a seven month research project at the Departement Communications, Telecom Paris. In 1989, he obtained the degree Master of Science in Digital Systems at Brunel University, London, U.K. His main areas of interest are protocol engineering for ATM networks, high performance protocol implementations, and performance evaluation

WAI CHEN received the B.S. from Zhejiang University, Zhejiang, China; and the M.S., M.Phil., and Ph.D. degrees from Columbia University, New York; all in electrical engineering. From 1984 to 1989, he was a graduate research assistant in the Center for Telecommunications Research (CTR) at

Notes on Contributors

Columbia University, where he carried out research in the design and evaluation of integrated services protocols. He joined SBC Technology Resources in 1989, where he was a senior technologist in the Advanced Network Technology group responsible for the analysis of the next-generation switching systems. Since 1992, he has been a **Research Scientist** in the Applied Research organization of Bell Communications Research (Bellcore). He has been, since March 1993, Bellcore's **Principal Investigator (PI)** of an ARPA-funded project on bandwidth management and access control for B-ISDN. His current research interests include resource management of wireless ATM access networks, mobile connection admission and dynamic traffic control in mixed wireless and wireline networks, and B-ISDN traffic management. He is a member of the IEEE, sigma Xi, and the Mathematical Association of America.

GIOVANNI CHIOLA received his degree of Doctor in Electronics Engineering from the Polytechnic of Torino in 1983. In 1986 he joined the Department of Computer Science of the University of Torino as a Researcher. In 1992 he became an Associate Professor. Since 1994 he is **Full Professor** at the Computer Science Department of the University of Genova. He (co)-authored more than 65 published papers in refereed journals and conferences. He has been appointed in the scientific program committee of 20 international conferences on Petri nets, performance modelling, simulation, and distributed systems.

MARCO CONTI received the Laurea degree in Computer Science at the University of Pisa, Italy, in 1987. In the same year he joined the **Networks and Distributed Systems department at CNUCE**, an institute of the Italian National Research Council (CNR). From 1989 to 1993 he was involved in a five-year national project "Progetto Finalizzato Telecomunicationi" aimed at designing and tuning the Italian broadband network infrastructure. He has worked on modeling and performance evaluation of Metropolitan Area Network MAC protocols. He has written over 50 research papers in the areas of design, modeling and performance evaluation of computer communication systems. His current research interest includes design, modeling and performance evaluation of ATM and Wireless Networks.

LORENZO DONATIELLO received the Laurea degree in Computer Science from the University of Pisa in 1978. From 1980 to 1990 he was with the Department of Computer Science, University of Pisa and since November 1990 he has been a **Professor** at the Laboratory for Computer

Science, University of Bologna. From April 1983 to October 1984 and from August 1986 to October 1986, he was visiting scientist at the IBM T.J. Watson Research Center, Yorktown Heights, NY. His research interests are in the area of performance and dependability evaluation of computer and communication systems, distributed simulation and performance models of wireless and ATM networks.

ASHOK ERRAMILLI obtained the B.Sc. degree in Physics from the University of Poona, India, the B.E. degree in Electrical Communications Engineering from the Indian Institute of Science, Bangalore, and the M.S. and Ph.D. degrees in Electrical Engineering from Yale University. He is **Director** of the Network Design and Traffic Analysis Research Group in Bellcore. His current research interests are in the analysis and control of fractal network traffic flows, broadband traffic management methods and traffic engineering economics.

ROSSANO GAETA was born in Torino, Italy in 1968. He received his doctor degree in computer science from the University of Torino in 1992. He is currently a **Ph.D. student** at the Computer Science Department of the University of Torino. His research interests include discrete event simulation of high level Petri nets, performance evaluation of parallel architectures, communication protocols and ATM systems.

EROL GELENBE was elected a Fellow of the IEEE at the age of 41 for "leadership in the development of computer performance evaluation". He has co-authored or authored three books with John Wiley and Sons and Academic Press in English, French and Japanese which deal directly with that area.

He received his B.S. in Electrical Engineering from the Middle East Technical University in Ankara (Turkey). He also holds the Master's and Ph.D. degrees in EE from the Polytechnic Institute of Brooklyn and the D.Sc. degree in Applied Mathematics from the University of Paris VI. He has been a chaired professor at the University of Liege (Belgium), and a Professor at the University of Paris.
In March 1993 he joined Duke University, and is currently the **Nello L. Teer Jr. Professor of Electrical and Computer Engineering, and ECE Department Chair** at Duke. His interests span Computer-Communication Networks and Distributed Systems, Neural Networks and their Applications, Learning Theory, and Discrete Stochastic Processes and

Notes on Contributors

Queuing Theory. He is the author of four books and of more than eighty journal articles, and of numerous conference publications.

His honors include the French Order of Merit (1992), the IFIP Silver Core (1980), the Parlar Foundation Science Award of Turkey (1994), and the France-Telecom Prize of the French Academy of Science (1996).

Dr. Gelenbe has supervised more than forty Ph.D. dissertations, and his former doctoral students are faculty members and senior researchers at many universities and research institutes in France, including University of Paris I, Paris VI, Paris IX, Paris XIII, Grenoble, Amiens, Rheims, Clermont-Ferrand, and Versailles, as well as at INRIA and Institut National des Telecommunications. Other former students are faculty members at the Technical University of Athens and the Middle East Technical University in Ankara (Turkey), and research staff members at IBM, AT&T, and other industrial organizations. He was the founding Vice-Chair of IFIP WG 7.3 in 1976, and then also chaired the organization for a subsequent term. His current funded research work includes automatic target recognition, distributed systems, and ATM networks. He is on the editorial board of several journals.

RICHARD GOLDING is a **Member of Technical Staff** at Hewlett-Packard Laboratories. He received his B.S. degree is Computer Science from Western Washington University in 1987, and the M.S. and Ph.D. degrees in Computer and Information Sciences from the University of California, Santa Cruz, in 1990 and 1991 respectively. His current research interests includes distributed operating systems, self-managing storage systems, fault tolerance, and wide-area collaborative applications.

DAVID HUTCHISON is **Professor** in Computing at Lancaster University and has been actively involved in research in local area network architecture and distributed systems for the past twelve years, first at Strathclyde University, then (since 1984) at Lancaster. He has completed many UK EPSRC-supported research contracts (including Alvey, ACME and Teaching Company projects) and published over eighty papers and a book in the area.

The main theme of his current research is architecture, services and protocols for distributed multimedia systems. He is involved in several recently-started projects in these areas, in which an integrating theme is Quality of Service for multimedia communications. Professor Hutchison is

Honorary Editor of the recently-launched international Distributed Systems Engineering Journal, published by the IEE, BCS and IoP. He is a programme committee member for many international workshops and conferences.

He is the UK representative on the Management Committees of the European Union COST14(CO-TECH) and COST237 (Multimedia Telecommunications Services) projects. He is active in standardisation, in particular in the UK working groups on OSI Quality of Service Framework and Enhanced Communications Transport Services.

GIANCARLO IANNIZZOTTO received the Electronics Engineering degree in 1988 from the University of Catania, Italy, in 1994, and is currently **a Ph.D student** in Computer Science at the same University. Since 1994 he has been engaged in research on Computer Vision and on Multimedia Systems with the Institute of Computer Science and Telecommunications of Catania and with the Faculty of Engineering of Messina. His current research interests include computer vision, image analysis, multimedia computing, video on demand and content-based retrieval in multimedia databases.

DEMETRES KOUVATSOS received the BSc degree in Mathematics from Athens National University in 1970, the MSc degree in Statistics from Victoria University of Manchester in 1971 and the PhD degree in Computation from UMIST, Institute of Science and Technology, University of Manchester in 1974. He is currently a **Reader** in Computer Systems Modelling, Department of Computing, University of Bradford. Since early 80's, he has been involved in research relating to queueing theory, stochastic modelling and systems performance evaluation. Dr. Kouvatsos has pioneered new and cost-effective methodologies for the approximate analysis of arbitrary queueing network models of computer and communication systems. The methodologies are based on the information theoretic principles of maximum entropy and minimum relative entropy, asymptomatic approximation techniques, queueing theoretic concepts and batch renewal processes. He has directed several EPSRC (Engineering and Physical Sciences Research Council, UK) and industrial research projects and is the author or co-author of many papers in the field. His proposed book entitled "Entropy Maximisation, Queueing and Computer Systems Applications" is in the final stages of preparation and will appear in 1997. Dr. Kouvatsos was the Chairman of the 6th UK Computer and Telecommunications Performance Engineering Workshop of

British Computer Society (BCS) in 1990, and also the Chairman of the first four IFIP Workshops on Performance Modelling and Evaluation of ATM Networks (1993-96). He was also the Co-Chairman of the 3rd International Workshop in"Queueing Networks with Finite Capacity" (1995). His latest technical interests include performance modelling and analysis of multiprocessor and database systems, B-ISDNs and discrete-time queueing networks. His professional associations include memberships with the IFIP Working Group WG 6.3 on the Performance of Computer Networks, the Hellenic Computer Society and the Performance Engineering Group of BCS and a Fellowship with the Royal Statistical Society, UK.

YI-BING LIN received his BSEE degree from National Cheng Kung University in 1983, and his Ph.D. degree in Computer Science from the University of Washington in 1990. Between 1990 and 1995, he was with the Applied Research Area at Bell Communications Research (Bellcore), Morristown, NJ. In 1995, he was appointed **Full Professor** of Department of Computer Science and Information Engineering, National Chiao Tung University. His current research interests include design and analysis of personal communications services network, mobile computing, distributed simulation, and performance modeling. He is an associate editor of the ACM Transactions on Modeling and Computer Simulation, a subject area editor of the Journal of Parallel and Distributed Computing, an associate editor of the International Journal in Computer Simulation, an associate editor of SIMULATION magazine, a columnist of ACM Simulation Digest, a member of the editorial board of International Journal of Communications, a member of the editorial board of Computer Simulation Modeling and Analysis, Program Chair for the 8th Workshop on Distributed and Parallel Simulation, General Chair for the 9th Workshop on Distributed and Parallel Simulation. Program Chair for the 2nd International Mobile Computing Conference, the publicity chair of ACM Sigmobile, Guest Editor for the ACM/Baltzer WINET special issue on Personal Communications, and Guest Editor for IEEE Transactions on Computers special issue on Mobile Computing.

DARRELL D.E. LONG is **Associate Professor** of Computer Science at the University of California, Santa Cruz. He received his B.S. degree in Computer Science from San Diego State University in 1984, and his M.S. and Ph.D. degrees in Computer Science and Engineering from the University of California, San Diego in 1986 and 1988 respectively.

His research interests include distributed systems, particularly high speed storage systems, fault tolerance, performance evaluation and mobile computing. He directs the Concurrent Systems Laboratory in the Baskin Center for Computer Engineering and Information Sciences.

He is a member of the Association for Computing Machinery, and a Senior Member of the IEEE Computer Society, serving as the Chair of the Technical Committee on Operating Systems.

YEN-WEN LU is a **PhD candidate** and **Research Assistant** in the Department of Electrical Engineering, Stanford University. His current research interests include parallel data routing, parallel signal/image processing, and low power VLSI design. He received his BS from National Taiwan University in 1991 and his MS in Electrical Engineering from Stanford University in 1993.

BISWANATH MUKHERJEE received the B.Tech. (Hons) degree from Indian Institute of Technology, Kharagpur (India) in 1980 and the Ph.D. degree from University of Washington, Seattle, in June 1987. At Washington, he held a GTE Teaching Fellowship and a General Electric Foundation Fellowship. In July 1987, he joined the University of California, Davis, where he became **Professor of Computer Science** in July 1995. He is co-winner of paper awards presented at the 1991 and the 1994 National Computer Security Conferences. He serves on the editorial boards of the IEEE/ACM Transactions on Networking and Journal of High-Speed Networks. He also served as Technical Program Chair of the IEEE Infocom'96 Conference. His research interests include lightwave networks, high-speed local and metropolitan area networks, wireless networks, and network security.

ADRIAN POPESCU received two PhD degrees in electrical engineering one from the Polytechnical Institute of Bucharest, Romania, in 1985 and another from the Royal Institute of Technology, Stockholm, Sweden in 1994. He is an **Associate Professor** in the Department of Telecommunications and Mathematics, University of Karlskrona/Ronneby, Sweden. His current research interests include B-ISDN/ATM networks, communication architectures and protocols, teletraffic theory, very high speed optical networks and LANs as well as dimensioning and optimizations of future integrated networks. He is a member of the IEEE and IEEE CS. His Email address is: adrian@itm.hk-r.se.

Notes on Contributors

ANTONIO PULIAFITO received the electrical engineering degree in 1988 from the University of Catania and the Ph.D. degree in 1993 in computer engineering, from the University of Palermo. Since 1988 he has been engaged in research on parallel and distributed systems with the Institute of Computer Science and Telecommunications of Catania University, where he is currently an **Assistant Professor** of computer engineering. His interests include performance and reliability modeling of parallel and distributed systems, networking and multimedia. During 1994-1995 he spent 12 months as visiting professor at the Department of Electrical Engineering of Duke University, North Carolina - USA, where he was involved in research on advanced analytical modelling techniques. Dr. Puliafito is co-author (with R. Sahner and Kishor S. Trivedi) of the text entitled *Performance and Reliability Analysis of Computer Systems: An Example-Based Approach Using the SHARPE Software Package*, edited by Kluwer Academic Publishers.

SALVATORE RICCOBENE received the electrical engineering degree from the University of Catania (Italy) in 1992. He is currently **completing** his studies toward the **Ph.D degree** in computer sciences at the same University. His interests include performance evaluation, parallelization of I/O subsystems, distributed systems, multimedia computing and intelligent agents for information retrieval.

JOCHEN H. SCHILLER is a **Research Assistant** at the Institute of Telematics, University of Karlsruhe. He received the diploma degree in 1993 and his doctoral degree in 1996, both from the university of Karlsruhe in computer science. His main interests include design, simulation, and implementation of high-performance communication subsystems, ATM-field trials with multimedia applications and ATM-hardware, FPGA and ASIC-based hardware design for time-critical protocol processing, and the use of different design methods for the construction of correct systems. Dr. Schiller is a member of IEEE and GI.

MATTEO SERENO was born in Nocera Inferiore, Italy. He received his Doctor degree in Computer Science from the University of Salerno, Italy, in 1987, and his Ph.D. degree in Computer Science from the University of Torino, Italy, in 1992. Since November 1992, he is a **Researcher** at the Computer Science Department of the University of Torino. His current research interests are in the area of performance evaluation of computer systems, modelling of communication networks and parallel architectures, queueing network and stochastic Petri net models.

Notes on Contributors

LORENZO VITA (M'88) received the electrical engineering degree from the University of Catania in 1979. From 1988 to 1994 he was an associate professor at the Institute of Computer Science and Telecommunications of Catania University. Presently, he is a **Full Professor** of computer science at the University of Messina, Italy. His research interests include the development of new computer organizations and architecture, parallel processing, and distributed systems. Dr. Vita is a member of IEEE society.

JONATHAN L. WANG received his B.S. degree from National Taiwan University, Taipei, Taiwan in 1981; and the M.S. and Ph.D. degrees from the University of Southern California, Los Angeles, CA in 1984 and 1988, respectively, all in Electrical Engineering. He is a **Principle Performance Analyst** in the Engineering, Performance and Control Department at Bellcore, where he works on traffic characterization, performance modeling and engineering support of a variety of communications services and technologies with a current focus on fast packet and broadband. He has an on-going dialog with network managers and traffic engineers from numerous network service providers and vendor companies.

WALTER WILLINGER received the Diplom (Dipl. Math.) from the ETH Zürich, Switzerland, and the M.S. and Ph.D. degrees from the School of ORIE, Cornell University, Ithaca, NY. He is a **Member of Technical Staff** at Bellcore, where he works in the Network Design and Traffic Analysis Research Group. He has been a **leader** of the work on self-similar traffic modeling at Bellcore and is co-recipient of the 1996 IEEE W.R.G. Baker Prize Award and the 1994 W.R. Bennett Prize Paper Award from the IEEE Communications Society for the paper titled "On the Self-Similar Nature of Ethernet Traffic".

GEORGE W. ZOBRIST received his BS and PhD in Electrical Engineering from the University of Missouri-Columbia in 1958 and 1965, respectively and his MSEE from the University of Wichita in 1961.
He has been employed by industry, government laboratories and various Universities during his career. He is presently **Chairman/Professor** of Computer Science at the University of Missouri-Rolla.His current research interests include: Simulation, Computer Aided Analysis and Design, Software Engineering and Local Area Network Design. He is presently Editor of IEEE Potentials Magazine, VLSI Design and International Journal in Computer Simulation.

Author Index

Rassul Ayani, 47

Frank Ball, 97

Michael S. Borella, 143

Georg Carle, 203

Giovanni Chiola, 33

Marco Conti, 117

Ashok Erramilli, 69

Rossano Gaeta, 33

Erol Gelenbe, xi

Richard Golding, 161

David Hutchison, 97

Giancarlo Iannizzotto, 187

Demtres Kouvastos, 97

Yi-Bing Lin, 223, 241

Darrell D. E. Long, 161

Yen-Wen Lu, 1

Biswanath Mukherjee, 143

Adrian Popescu, 47

Antonio Puliafito, 187

Salvatore Riccobene, 187

Matteo Sereno, 33

Lorenzo Vita, 187

Jonathan L. Wang, 69

Walter Willinger, 69

Subject Index

Aggregate traffic, 79
Asynchronous updates, 162
ATM, 98, 102, 103, 104, 113
ATM networks, 203

Bandwidth allocation, 106, 113
Buffer ineffectiveness, 83
Buffer management, 70
Bursty traffic, 71
Busy line, 259

Cell, 223
Channel configuration, 6
Chaotic map, 79
Circuit switching, 48
Coloured Petri nets, 36, 38, 39, 40
Communication Protocol, 34, 35, 36
Connection admission control, 70
Continuous media, 97, 187
Controls, 80
Credit feedback, 3

Discrete Event Simulation (DES):
    Asynchronous timing, 124
    Output analysis, 125-127
    Simulator design, 123-125
    Steady-state simulation, 126-127
    Synchronous timing, 124
    Terminating simulation, 125
    Transient simulation, 125-126

Disk array, 188,197
Distributed Queue Dual Bus (DQDB):
    Bandwidth balancing mechanism (BWB), 132
    Performance evaluation, 129-134
    Protocol, 128-129
    Simulation, 130

Elevator polling,
Enhanced Network Layer Architecture (ENLA), 97
Equivalent bandwidth approach, 86
Erlang-B formula, 70
Error correction, 204
Explicit de-registration, 228

Fast simulation of rare events, 140
FBM: fractional Brownian motion, 74
FDDI, 98, 101, 103, 104, 106, 110, 111, 113
Finite-buffer model,
Finite buffer simulation, 82
Fractal, 69
Fractional ARIMA model, 91

Generalised Exponetial(GE) distribution, 98, 107
GE-Type queueing model, 98, 108, 109, 110, 113, 115, 116
Group communication, 165, 204

Hand-off, 234, 242

Heavy-tailed distribution, 71
Heterogeneous system,
HLR: Home Location Register, 223
Hurst parameter, 75
Hybrid simulation, 140

IDA, 190,199
IDC: index of dispersion of counts, 70
IEEE 802.6 standard, 128
Independent replications, 126-127:
　　Confidence interval, 127
　　Confidence level, 127
Infinite variance, 71
Isochronous traffic, 48

Joseph Effect, 71

Local Area Network (LAN), 117-119
Local Area Networks, 47
Location forwarding, 233
Long holding times, 90
Low frequency, 75
LRD: long-range dependence, 71

Mandelbrot's construction, 76
M-ES: mobile end station, 241
MCN: mobile computing network, 241
MDBS: mobile data base stations, 241
Media Access Control, 47
Medium Access Protocol (MAC): 117-119, 128-129, 137
　　Performance measures, 118-119
Metropolitan Area Network (MAN), 117-119, 128
Mobile computing network, 241

Mobile data base stations, 241
Mobile end station, 241
Multimedia, 187,193
Multiplexing gains, 70

Networks,
Network traffic, 176
Noah Effect, 71

OSS: operations support system, 88

Packet train model, 79
Parallel and Distributed Simulation (PADS), 140
Parsimony, 74
PCS subscriber, 223
PCS: personal communications services, 223, 242
Peakedness, 70
Persistence, 71
Policing, 87
Portable, 242
Protocol processing 210

Quality of Service (QoS), 81, 97, 106
Quality of Service Architecture (QoS-A), 97
Queue length distribution, 81

RA: registration area, 223
Registration area, 223
Response time distribution, 98, 107, 109, 110, 113, 116
Resource reservation, 97
Request buffering, 241
Request waiting, 241, 250
RMD: random midpoint displacement method, 76

Safe operating points, 73

Subject Index

Scaling, 76
SCAN polling,
SDL, 207
Self-similar, 69
Shaping, 87
Shuffling, 72
Simulation, 39, 40, 41
Stochastic activity
Stochastic Well-formed Nets, 36, 38, 39, 40
SVC: switched virtual connection, 89

Temporal mapping, 98, 99, 103, 106, 110, 113
Temporary local directory number, 227
Timeout de-registration, 228
Token-exchange, 7
Traffic management, 70
Traffic measurements, 70

User re-dialing, 241

VBR: variable-bit rate, 91
Variable Bit Rate (VBR) traffic:135-136
    Bandwidth allocation, 135-137
VHDL, 210
Video on demand, 187,190

Visitor Loaction Registers, 223
VLR: Visitor Location Register, 223

WACS: Wireless Access Communication Systems, 228
Wavefront arbitration, 5

Wavelength dedicated to application, 48
Wavelength Division Multiplexing, 48
Weak consistency, 161
Wide-area services, 161
Wireless Access Communication Systems, 228
Wireless Local Area Networks (WLAN): 117, 137-139
    Packet Reservation Multiple Access (PRMA), 137-139
Workload characterization, 120-123
    Markovian model, 122-123
    Synthetic, 120-121
    Trace, 121-122